Ernst Schering Research Foundation Workshop
Supplement 5
Interferon: The Dawn of Recombinant Protein Drugs

Springer-Verlag Berlin Heidelberg GmbH

Ernst Schering Research Foundation Workshop
Supplement 5

Interferon: The Dawn of Recombinant Protein Drugs

J. Lindenmann, W.-D. Schleuning
Editors

With 7 Figures and 1 Table

Springer

Series Editors: G. Stock and M. Lessl

ISSN 0947-6075
ISBN 978-3-662-03789-8

CIP data applied for

Die Deutsche Bibliothek – CIP-Einheitsaufnahme
Schering-Forschungsgesellschaft <Berlin>: [Ernst Schering Research Foundation Workshop / Supplement] Ernst Schering Research Foundation Workshop. Supplement.
Reihe Supplemnt zu: Schering-Forschungsgesellschaft <Berlin>: Ernst Schering Research Foundation Workshop
5. Interferon.. - 1999
Interferon: the dawn of recombinant protein drugs / J.Lindemann ; W.-D. Schleuning ed.
(Ernst Schering Research Foundation Workshop : Supplement ; 5)
ISBN 978-3-662-03789-8   ISBN 978-3-662-03787-4 (eBook)
DOI 10.1007/978-3-662-03787-4

Typesetting: Data conversion by Springer-Verlag

SPIN:10730445   13/3134–5 4 3 2 1 0 – Printed on acid-free paper

Preface

Forty years of Interferon

I wish to dedicate this short introduction to the memory of Alick Isaacs (1921–1967), and to that of Sir Christopher Andrewes (1896–1988).

Let us go back more than 40 years. In 1956 Isaacs was in charge of the World Influenza Centre. Andrewes was head of the division of bacteriology and virology, and deputy director of the National Institute for Medical Research in London.

When researchers are faced with a seemingly new phenomenon, explanations are easy to come by. These explanations fall into two broad categories: the phenomenon in question is either due to something or to the lack of something. I apologize for the primitive way in which I express this, but I am going to give three examples, scattered over 100 years, of what I mean.

First example: in 1880 the great French microbiologist Louis Pasteur was involved in work on chicken cholera. He was struck by the following observation: if a suitable chicken broth was inoculated with the bacterium, the organism grew profusely and the liquid became turbid. If he now freed the fluid, by sedimentation or filtration, from the bulk of the organisms and re-inoculated it with the same bacterium, no growth occurred. He immediately offered two possible explanations: either the bacterium, during its first round of growth, had used up some essential constituent of the broth or it had produced a substance which,

The participants of the workshop

upon reaching a certain concentration threshold, inhibited further growth. This second alternative would have been analogous to alcoholic fermentation, with which Pasteur was familiar. In the case of chicken cholera Pasteur attempted to but failed to reveal the existence of such a hypothetical inhibitor. He therefore settled for the first explanation, the so-called exhaustion hypothesis, which he extended to the phenomenon of immunity in vivo. In the following years a lively debate developed over the issue of whether immunity was caused by the lack of something or by the addition of something.

Second example: 100 years later, in a paper in Lancet by Murray, Murray, Murray and Murray (1980) on the rarity of planar warts in Cushite nomads, the authors wrote: "There may be a factor ... which inhibits virus replication. Alternatively, there may be a deficiency ... of a nutrient essential for the replication of the virus."

Third example, and now I am getting closer to my subject: almost halfway between the first two examples, in 1942, Christopher Andrewes published an important paper on viral interference in tissue culture. In this he stated: "The virus first upon the scene uses up some essential foodstuff in the cell. An alternative ... hypothesis ... would be, of course, the generation of some inhibitory substance."

Notice the words "of course". They show Andrewes' modesty and generosity. To him it was clear that anybody who gave only 5 min thought to the problem would hit upon these two explanations. Like Pasteur 60 years earlier he attempted to show the existence of such an inhibitor, but without success.

There the matter rested when Alick and I started experiments on viral interference, using an extraordinarily simple system which allowed repeated manipulations of fragments of chick chorioallantoic membranes suspended in a salt and glucose solution. We had one advantage over what Andrewes had done earlier: whereas he had used interference between two strains of live virus, it had become clear in the meantime that interference also occurred between inactivated virus and live virus. The use of inactivated virus as the interfering agent made it from the start less likely that the explanation which Andrewes had favored, the exhaustion of some essential foodstuff, also applied in our case, although it was by no means excluded, since the inactivated virus, after all, did something. In fact, we entertained for a time the idea that some sort of abortive replication of the interfering virus occurred.

Alick once told me that when trying to explain to one of his children what he did, he had said: "Well, you see, we ask questions and then we try to answer them." Whereupon his son said: "And they pay you for that?"

What were the questions that we asked? In fact, they had little to do with interference per se. Before coming to Mill Hill I had done a few experiments on viral interference in which I had used inactivated influenza virus firmly attached to red cells as the interfering agent. This technique appealed to Alick because he thought that it might be used to answer the following question: does the virus, as was known at that time from bacteriophages, inject its nucleic acid into the host cell, the viral envelope remaining outside? Or does the whole virus enter the cell? He proposed to use red cell ghosts (I had used intact red cells) coated with virus particles and to look at them with the electron microscope before and after they had induced interference. To do this he enlisted the help of an excellent electron microscopist, Robin Valentine.

This approach proved impracticable. The virus particles were clearly visible at the beginning, but after the coated ghosts had induced interference, the picture became blurred by cellular debris and could not be interpreted.

I believe this is a clear example of how experimental science is being done. Philosophers of science seem to believe that scientists first ask questions and then look around for experimental techniques that might answer them. I suspect it is the other way round: Scientists hit upon some nice technique and then they shop around for questions that could be answered. To put it aphoristically: Scientists don't do experiments in order to answer questions, they ask questions that allow them to do experiments.

In the course of this work we made an unexpected observation: when we tried to exhaust the interfering power of virus-coated ghosts by repeated rounds of interference, it looked as if new interfering activity was being generated. I called this hypothetical "something", jokingly at first, interferon, and the first experiment that addressed this new question is labeled in Alick's lab notebook, now at the National Library of Medicine in Bethesda, "In search of an interferon".

In search of an interferon does not mean that we were expecting several interferons, but rather, that we were looking for anything that would fit the notion that something was there and, of course, as Andrewes might have said, we had to exclude the possibility that the phenomenon was caused by the absence of something

Figure I shows Alick in happy days with his twins. Somebody quipped: Alick always does his experiments in duplicate. This may have been a good joke, but it was not true. In fact, Alick told me he had learned from Sir Macfarlane Burnet that one should never repeat an experiment in exactly the same manner. Results of experiments should always be taken seriously. If you repeat an experiment painstakingly and then get the same result, you are no wiser than you were before. And should you get a different result, then both experiments cancel each other. So we did a series of experiments, the last one always overlapping the precedent one to some extent, but always with something new added. For instance, we titrated the activity, we attempted to pass it through filters, to centrifuge it, to neutralize it with antiviral antibodies etc. At the beginning of 1957 we were convinced

Fig. I. Alick Isaacs with his twins

that something was there, not the lack of something. People who were skeptical, as of course one should be, called it a misinterpreton.

I left Mill Hill in 1957, my fellowship having expired. I was anxious to find a research object that would be independent of interferon. With many others I shared the optimism that within a short time the purification of interferon, a biochemical business in which I did not want to engage, would be achieved, and that then biological questions could be asked more meaningfully. In the meantime I wanted to do experiments. An opportunity presented itself when by chance I observed an inbred mouse strain that was highly resistant to orthomyxoviruses. I wrote about this to Alick, and in his reply he thought it unlikely that this had anything to do with interferon.

This was perfectly reasonable advice, because interferon was thought to inhibit many different viruses, whereas the resistance phenomenon I had observed was limited to a few related viruses. Many experiments suggested themselves that were easy to perform, and again the question had to be asked: is it something or the lack of something? During the following 18 years several collaborators and myself looked at the presence of inhibitors, at the absence of receptors, at powerful immune responses or the lack of immunopathology. All these experiments were fun to do, but gave negative results. The resistance phenomenon remained as unexplained as ever.

The resistance proved to be caused by a single dominant gene, which we called Mx. In the course of the years we had developed, again one of those experiments that suggested themselves, so-called congenic lines of mice. By systematic back-crosses over some 20 generations we had obtained two lines of inbred mice which were genetically identical except for a small segment of DNA containing the Mx locus. We argued that this locus yielded a gene product, a protein, which should be present in the resistant line but absent in the susceptible line. Because the two lines were congenic, this was expected to be the only substantial difference, and hence it should be possible, by immunizing susceptible mice with tissue extracts from resistant mice, to get an antiserum capable of recognizing the Mx gene product.

We did this carefully, and tested the putative antiserum by the methods then available, by precipitation, gel diffusion, immunofluorescence, complement fixation, neutralisation, later by Western blotting, all without success. And then there was a sudden burst of light. My collaborator Otto Haller obtained from Ion Gresser a small sample of an extraordinarily potent anti-interferon serum, and a new experiment became possible: when resistant animals were treated with this antiserum, they became fully susceptible.

Now it dawned upon us: the Mx gene product is normally switched off, and interferon switches it on. It was interferon after all, but acting indirectly through the Mx gene. We could now repeat the immunization experiment more intelligently, by injecting, into susceptible mice, not simply organ extracts of resistant animals, but of interferon-treated resistant animals. Using this approach, highly specific antibodies were readily obtained and proved useful for the cloning of Mx.

So resistance is caused by something, the presence of an interferon-inducible gene. But we could have asked the other way round: what is the cause of susceptibility? And here the answer would have been: it is the lack of something, the lack of a functional Mx gene.

Of course, as Sir Christopher would have said.

J. Lindenmann

Table of Contents

List of Editors and Contributors

Editors

W.D.Schleuning
Schering AG, D-13342 Berlin, Germany

J. Lindenmann
Obere Geerenstrasse 34, CH-8044 Gockhausen, Switzerland

Contributors

D. Burke
13 Pretoria Road, Cambridge CB4 1HD, UK

K. Cantell
Vänrikki Stoolin katu 9 A 9, SF-00100 Helsinki, Finland

N.B. Finter
9 Burntwood Road, Sevenoaks, Kent TN13 1PS, UK

P. Fitzgerald-Bocarsly
Department of Pathology and Laboratory Medicine, UMDNJ, New Jersey Medical School, 185 So. Orange Avenue, Newark, NJ 07103, USA

I. Gresser
74 avenue de Villiers, F-75017 Paris, France

O. Haller
Institut für Medizinische Mikrobiologie und Hygiene, Abteilung für Virologie, Hermann-Herder-Straße 11, D-79104 Freiburg i. Breisgau, Germany

Y. Kawade
43-6 Okazaki Minamigosho-cho, Sakyo-ku, Kyoto 606, Japan

J. Lindenmann
Obere Geerenstraße 34, CH-8044 Gockhausen, Switzerland

I. Löwy
U-158. INSERM, Hopital Necker-Enfants Malades, 149 rue de Sèvres, F-75749 Paris, France

H.F. McFarland
Neuroimmunology Branch, National Institutes of Health, National Institute of Neurological Disorders and Stroke, Building 10, Room 5B16, Bethesda, MD 20892, USA

T. Pieters
Vrije Universiteit Amsterdam, School of Medicine, Section Medical History, Van der Boechorststraat 7, NL-1081 BT Amsterdam, The Netherlands

W.-D. Schleuning
Schering AG, D-13342 Berlin, Germany

P. von Wussow
Hannover-Zentrum, Dieterichstraße 35b, D-30159 Hannover, Germany

1 Is There Life Without Interferon?

N. B. Finter

I have chosen as my title "Is there life without interferon?" for two reasons. First, the evidence suggests that the interferons (IFN) appeared very early in evolutionary time, and so early that one can reasonably suppose they have a very fundamental role in multicellular organisms. Figure 1 is adapted from one in a 1985 paper by Weissmann and colleagues [1], and shows that the primitive IFN-α and IFN-β genes had already separated many million years ago (MYA). Roberts and colleagues [2] suggest that the best estimate for the time of this is approximately 235 MYA. So the primordial IFN-α/β gene is older, and possibly very much older. Indeed one can speculate that it may have evolved at the time when the loose association of single cells led to the first primitive multicellular organisms, and served to prevent the nucleic acid from one cell pre-empting the functions of another.

As shown in Fig. 1, other gene families have since separated from the primitive IFN-α gene. The omega genes diverged about 130 MYA, and shortly before the start of mammalian evolution. In turn, the tau genes separated from an IFN-ω gene about 36 MYA and are present in certain members of the sub-order *Artiodactyla*, including the sheep and cow: their protein products play an essential part in maintaining embryo survival in the early stages of pregnancy. Another family, the delta genes, which also diverged from IFN-α before mammals evolved (not shown), has as yet only been identified in the pig, and its role is as yet unknown. Finally, as the result of duplications, cross-over and recombination events, the single IFN-α gene has given rise to more and more variant genes during the 100 million year course of mammalian evolu-

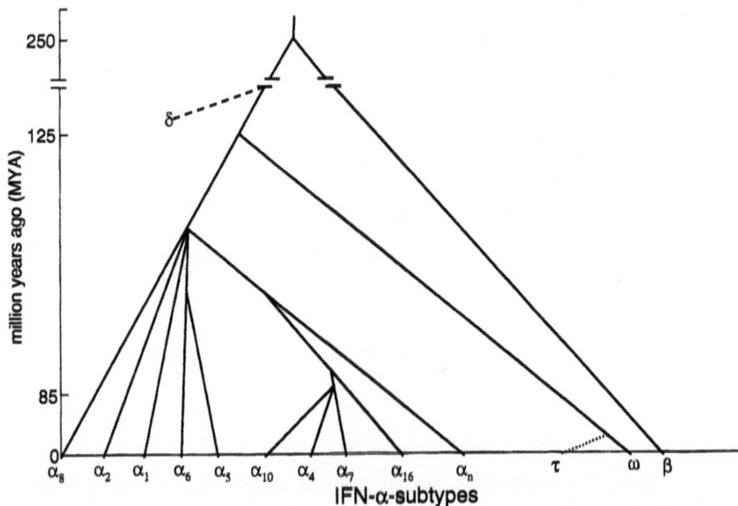

Fig. 1. Postulated evolution of the human interferons (*IFN*). Nine of the 13 IFN-α are shown. α_n: subtype not identified (data from[1]). The IFN-α and IFN-β pathways diverged about 250 million years ago (MYA), after the separation of the birds [2]. The gene for IFN-δ arose from the IFN-α lineage an estimated 180 MYA; it has only been found in the pig, but may be represented in other mammals. The IFN-τ diverged from an animal IFN-ω about 36 MYA and are found only in ruminants; there is no human IFN-τ [2]

tion. In man, there are 13 genes for IFN-α, all closely clustered on the short arm of chromosome 9. These give rise to chemically distinct interferon proteins, the subtypes of human IFN-α [3,4], and increasing evidence shows that several of these have specialised functions which are not shared by others, or are only shared when they are applied at a very much higher concentration [5–7].

From all this, it is apparent that over millions of years, the interferon system has become progressively more and more complex, and the individual interferon proteins more and more specialised. It seems that the interferons have a role in very many cellular activities, and rank among the most important mechanisms for maintaining cellular homeostasis. So indeed, there is no life without interferon.

But my title has another more facetious meaning. Having now worked with interferons for nearly 50 years, even if for 7 years without knowing it, no doubt like some others participating in this symposium, I find it hard to imagine what my life would have been if they had not been discovered. Probably I would have gone back to clinical medicine, for in October 1948, I had just completed 9 months as a junior doctor in the cardiac surgery group at Guy's Hospital, London, then the pioneer centre in Europe for open heart surgery. The work was enormously exciting, and it was a very difficult decision to give it up. But I had long wanted to try my hand at medical research, a desire spurred on by witnessing as a student the dramatic results obtained in the first trials of penicillin in civilians. Later, as a junior doctor, I had care of two patients enrolled in the Medical Research Council blind controlled trial of streptomycin in tuberculosis, which set the pattern for most clinical trials since. So even against advice from my surgical chief at Guy's, I decided to return to Cambridge, my old University, to take up a 5-year appointment as a junior lecturer in the Department of Pathology.

Soon after I arrived, one of the senior lecturers in the Department, Michael Stoker, suggested I might like to study the epidemiology of mumps, there being an outbreak in local children at the time. This seemed likely to be interesting, and so I started to look into the literature. This proved a shock. I had taken a special interest in microbiology during my time as a medical student and afterwards, but I knew very little about viruses other than that they were thought to be tiny bacteria, invisible under the light microscopes of those days, and apparently impossible to culture other than by animal inoculation. So I was astonished to come across an enormous textbook of virology, written by two eminent Toronto scientists, Rhodes and van Royen, and packed with information. I also found that techniques had been developed whereby many viruses, including myxoviruses, such as influenza and mumps, could be grown in the fetal membranes of developing chick embryos. Fertilised hen's eggs could now replace animals for the culture of such viruses, which immediately made their study very much easier. I read, I started work, and I became fascinated.

The work had a further attraction, for I was given a special allocation of 100 eggs a week for my virus culture studies. At that time, war-time food rationing had been continued in the UK, and each person received only one or two eggs a week. I soon discovered that when 100 eggs had been incubated for about 5 days, and were "candled", i.e. transilluminated with a strong beam of light, about 70 contained the living embryos needed for virus studies. Some 15–20 of the remainder proved not to have been fertilised, and although by then not exactly " new laid", were nevertheless edible, a welcome addition to my bachelor diet, and a very acceptable gift for my more attractive female acquaintances of the time!

I soon succeeded in isolating mumps virus from the saliva of patients, and injected eggs with very large amounts of these isolates in order to grow stocks of virus. When I tried to measure their content of infectious virus, I soon ran into problems, for the results often did not accord with a Poissonian distribution of infectious particles, and it was impossible to calculate a 50% infectivity titre in the conventional way. In a repeat assay, the results often proved equally erratic. Also, I found I was quite unable to repeat some of the experiments with mumps virus described in the literature by workers in Burnet's laboratory in Melbourne. I told Burnet of my difficulties when he happened to visit Cambridge, and his replies made it evident that he regarded me as totally incompetent (much later, I learned that the Australian work could only be repeated with one particular mumps strain). My colleagues could not help me and so I decided to write to Werner Henle, the other acknowledged expert on mumps virus at that time. In my letter, I also asked if there was any chance of a short-term position in his laboratory in Philadelphia. To my great surprise, he replied immediately to say he would be pleased to have me, Cambridge generously gave me a year's unpaid leave of absence, and so I moved to Philadelphia in September 1950.

Werner was the grandson of the great German medical scientist whose name is known to every doctor from his discovery of Henle's loop in the kidney. In view of his illustrious lineage, Werner might perhaps have escaped the fate of most Jews during the Nazi regime, but appalled by political developments, he and his wife, Gertrude, also a doctor, decided to emigrate to the United States before World War II. Their talents as virologists were soon well-recognised, and within a few years, Werner was made the Director of a virus laboratory in the Childrens Hospital of Philadelphia. This was housed in extremely cramped and uncomfortable conditions in an area of the basement of the old hospital at 1740 Bainbridge Street which had been declared unfit for use by the hospital laundry workers. The hospital itself was situated in a very delapidated and unsavoury down-town area. Nevertheless, or perhaps in consequence of these features, those working in the laboratory were full of enthusiasm and energy, and all worked very hard and for long hours. Werner was an excellent research director, always willing to discuss data and suggest new experiments, but allowing each individual much freedom. I very soon discovered that his main interest, and the main focus of the work in the laboratory, was the virus interference phenomenon, which he had long studied. With very little hesitation, I decided to shelve my plans to work on mumps virus and join in the studies.

The virus interference phenomenon was first described independently in 1935 by two scientists. In the USA, Hoskins [8] infected a total of 40 monkeys on 18 occasions with a neurotropic strain of yellow fever virus, and superinfected them with an otherwise lethal dose of a pantropic strain of the virus. He reported that 13 monkeys survived out of 18 if both viruses were given simultaneously, 2 monkeys died from yellow fever, and 3 died for other reasons. When the interval between the first and second infection was between 10 and 24 h, 12 monkeys survived out of 22; but if the interval was more than 24 h, only 2 survived out of 8. Magrassi [9] working in Italy with two strains of herpes virus, independently obtained essentially the same result. Neither could offer any explanation.

At that time, no virus disease could be prevented or treated, apart from the very few for which there were vaccines which were very primitive by today's standards. This was a matter of great concern, for memories of the great 1918–1919 influenza pandemic were still vivid. In this greatest public health disaster since the Black Death, an estimated 25 million people died world-wide without respect for age, health status, sex, race or geography. By October 1918, 20% of the US Army was ill with influenza, and 24,000 died; total battle casualties were only 10,000 more. Approximately 25% of the population of Samoa died. In view of such terrible statistics, it is not surprising that the description of the interference phenomenon aroused considerable interest: hopefully, if the underlying mechanism or mechanisms could be unravelled, perhaps some general way of controlling a virus infection might emerge.

At that time, when viruses could only be studied by using animals, usually rodents, guinea-pigs or monkeys, the work was very laborious and imprecise, and progress was correspondingly very slow. Nevertheless it was soon found that one virus might interfere with the growth of another even when the two were antigenically totally unrelated. In the 1940s, the introduction of the use of fertilised hens' eggs for virus culture greatly reduced the technical problems, but during these war years, most of the still very small number of virologists turned their attention from research to pressing short term problems. After 1945, work on the interference phenomenon was soon resumed and some progress was made. One major advance was the demonstration by Henle and Henle [10,11] that influenza virus could be irradiated with ultra-violet light until it could no longer replicate, but still retained its capacity to interfere with the growth of a range of viruses, whether related or unrelated. By eliminating one variable in a system which had previously involved two independently replicating viruses, this finding greatly facilitated studies on interference. Nevertheless the underlying mechanism(s) remained an enigma, and in 1950 in a lengthy review of work to that date, Henle [12] discussed a number of possibilities, but could not come to any definite conclusion.

Those who studied virus interference in Henle's laboratory in the early 1950s and workers in other laboratories must certainly have handled chick or other interferons during the course of their research, although without knowing it. Indeed in retrospect, there are at least two publications in which the authors seem to have clearly identified the effects of an interferon, but failed to appreciate the significance of their findings.

Lennette and Koprowski [13] studied the 17D vaccine strain of yellow fever virus and a number of antigenically related and unrelated viruses in primitive tissue mince cultures of mouse or chick embryos, and measured the amounts of virus formed by mouse inoculation. They noted that media or cell extracts made from cultures 24 h after exposure to 17D virus contained a non-viral substance that passed through the 12 µm pores of a filter membrane, possibly the first description of an interferon, but because their virus assay measurements were so imprecise, they discounted the result. They carried out further experiments in eggs, but overall, concluded that some fundamental mechanism common to all the systems studied must be responsible for interference, that the effects noted were "not attributable to a non-specific substance arising as the result of multiplication of the initially inoculated virus; but that "additional information was desirable".

Burnet and Fraser [14] got even nearer to discovering interferons. They reported that intravenous injection of the NWS strain of influenza virus into a developing chick embryo resulted in lethal haemorrhages in the brain. If however the chorio-allantoic membrane was simultaneously infected with the same virus, or with MEL, another strain of influenza virus, no brain haemorrhages resulted. They decided that this "was governed by some component other than virus in the circulating blood". This might result from a positive effect, some component of the virus-cell interaction in the membranes "passing to the vascular endothelium and so modifing its reactivity in the required fashion". Alternatively, the effect might be negative, with removal from the blood of some factor needed for virus multiplication in the brain. From their point of view very unfortunately, they rejected the first explanation in favour of the alternative hypothesis. How they must have kicked themselves later!

The Walter and Eliza Hall Institute in Melbourne was then one of the world's most active centres of virus research. Scientists there showed that exposure of chorio-allantoic membrane cells to one influenza virus could impede the subsequent attachment of a superinfecting dose of the same or some other strains of influenza virus and other myxoviruses, but not the attachment of others. For a time it seemed that this so-called receptor gradient might be the basis for at least some examples of virus interference. Amongst those who studied this idea was a young Scottish medical scientist, Alick Isaacs. He spent a year in Burnet's laboratory in 1956, and published several papers with Margaret Edney and other collaborators dealing with various aspects of the interference phenomenon. On his return to the United Kingdom, Isaacs was appointed head of the World Health Organisation Influenza Laboratory, which was housed in the National Institute for Medical Research (NIMR), the central laboratory of the Medical Research Council, in Mill Hill, a northern suburb of London. There he was joined in 1956 by a young Swiss medical virologist, Jean Lindenmann. Discussions with his laboratory head in Zurich, Hans Mooser, had given Lindenmann an idea for a new approach to the study of virus interference, and when he met Isaacs, the two decided to work together. Success comes to the prepared mind, and during the course of their joint research, which Pieters [15] has well documented, Isaacs and Lindenmann identified a protein mediator which was released from chick chorio-allantoic membrane cells treated with an interfering virus and appeared in the medium. They showed that the activity was due to a protein, and that when freed from any contamination with virus, was still able to reproduce the virus interference phenomenon in fresh chorio-allantoic membrane cells. They christened their factor "the interferon" [16]. A new branch of biological science was opened.

My work with the Henles did not help to unravel the mechanisms underlying virus interference, but led to the publication of three papers in the very prestigious Journal of Experimental Medicine. After my year in Philadelphia, I returned to complete my 5-year appointment at Cambridge, and in 1953 and largely on the strength of these papers, I was offered a 3-year appointment on the external staff of the Medical Research to work with Leslie Hoyle at Northampton.

This is not the place to write about Hoyle. Suffice it to say that he had one of the most brilliant and original minds that I have ever encountered, combined with a refreshing confidence in his own abilities and a total intolerance of authority. Not surprisingly, these qualities did not endear him to the leaders of the UK scientific establishment, few of whom were his intellectual equals, and he was effectively banished from a post in any of the major microbiological centres of the time. Fortunately, his capabilities were well appreciated by some, including Sir Graham Wilson, co-author with Topley of the great textbook of microbiology, which is still regularly updated and republished. Wilson was then the Director of the Public Health Laboratory Service, which consists of a chain of laboratories throughout the U.K. providing diagnostic and public health facilities for the surrounding area. Wilson appointed Hoyle to head the small laboratory in Northampton, a market town in the middle of England. This position gave him plenty of time to continue his research into the fundamental nature of animal viruses and their method of replication. Here, he was years ahead of his time in recognising and proving that animal viruses were fundamentally different from bacteria, and essentially the same as the bacterial viruses, the bacteriophages. that were already well-recognised.

Hoyle had no great interest in interference, and I soon found that I could contribute little or nothing to his research programme. So by mutual but unspoken agreement, we worked separately. I again had a supply of fertile eggs for my work, but my allocation of 100 eggs a week seemed pitifully small compared with the several thousand a week that had been available in Philadelphia. I realised I needed some alternative, and soon found that after a fertilised hen's egg had been incubated for 12–13 days, it became easy to open the end, and decant the embryo and yolk sac, leaving the outer layer of the chorio-allantoic membrane attached to the egg shell; this could be cut up into small squares. Each such piece could be used instead of a whole egg for virus infectivity titrations etc. To my great disappointment, the paper describing the technique, my first entirely independent publication, was published a few months after a very similar one by Fulton and Armitage [17], who had independently hit on the same idea. This latter publication had an important consequence, for it influenced Alick Isaacs to use chorio-allantoic membrane pieces in his interference studies. As Henle and Henle [18] later commented, this was probably crucial to the discovery of interferon.

I left Hoyle's laboratory in 1956 to join the staff of a new pharmaceutical company, a division of the British chemical giant, ICI. My brief was to look for chemicals with useful antiviral properties, but I was given almost total freedom to continue my research on interference. In 1959, I joined the scientists from two other pharmaceutical companies and workers from the NIMR in the so-called Scientific Committee on Interferon. This MRC Committee chaired by Isaacs, had the remit of expediting the development of interferons as agents for medicinal use. I have knowingly worked mainly on interferons from then till today.

I have to confess that when the paper describing "the interferon" was published, my initial reaction was one of great disappointment. Werner and Brigitte Henle were my scientific heroes, and it seemed very sad that after all their years of work on virus interference, they had not made the crucial discovery. At first, I managed to persuade myself that there were possible technical flaws in the published work, and that there might be other explanations for the data, but I soon changed my views. In retrospect, even though the Henles failed to discover the interferons, they had a very big even if indirect influence on their development. They stimulated interest in interferons in those such as Kari Cantell, Kurt Paucker, myself and others who had the good fortune to work in their laboratory, and through their later contributions, greatly hastened progress with interferons. Without these contributions, there would not have been the supply of human leucocyte interferon, meticulously though laboriously prepared in Finland, which was crucial for arousing and maintaining clinical interest in the 1970s; Ion Gresser's group might not have continued to explore interferons as anti-tumor agents without the demonstration that interferons slowed the growth of a mouse tumor cell in vitro, and might not have found a source of mouse interferon of sufficient potency to carry out their experiments; and human lymphoblastoid interferon would not have been developed. So when medical history is written, I hope that among his other achievements, Werner Henle's part in the development of the interferons will be recognised, and for this alone he will be considered at least as illustrious as his distinguished grandfather.

If Isaacs and Lindenmann had not discovered "the interferon" in 1957, how long would it have been before someone else did? During the1950s, Nagano and colleagues in Japan tried to determine why irradiated vaccinia virus interfered with the growth of active vaccinia virus in the skin of rabbits. In 1954, they reported that a virus-free soluble "Facteur Inhibiteur" extracted from the skin after injection of the inactivated virus was able to reproduce the interference effect [19]. In 1981, Nagano [20] summarised experiments over the years with this factor, which appeared to contain or be carbohydrate in nature. Perhaps his studies would eventually have led to the identification of an interferon, even if they had not been discovered in 1957, but he seems to have had very limited resources available, and the work must have been made extremely difficult because of the complex composition of the skin extracts used.

It seems more probable that the Virus-Inhibitory Factor (VIF) which Ho and Enders described in 1959 [21,22] would have stimulated general interest, and led to the interferons under this name. But they acknowledged in their papers that VIF was very similar to an interferon, and it is not now clear whether they would have undertaken their studies if interferon had not been discovered. Although as a very distinguished microbiologist and Nobel prize winner, John Enders had enormous prestige, his work with VIF might not have been followed up. Perhaps then the interferons would not have been identified until, many years later, when those working on the human genome would have identified a cluster of genes on the short arm of chromosome 7, and studied the functions of their products. In fact, a quite different approach would probably have led sooner to the interferons.

The 1950s were the golden era of antibiotics. In turn the introduction of penicillin, streptomycin, chloramphenicol and the tetracyclines had succeeded beyond wildest expectations in leading to the control of most bacterial infections. Only viruses remained as the uncontrollable causes of acute and chronic infectious disease. In those now distant days, the fundamental difference between viruses and bacteria was not appreciated by more than a handful of scientists, and the response of academic and industrial scientists to the challenge of virus infections was to test more and still more fungal materials for antiviral activity. As the result there were very many claims for success. All later proved to result from the toxic effects of the mould product on the cells used to grow the virus concerned, with two notable exceptions. One was the chance discovery made by Richard Shope from Yale University. Returning from a European lecture tour, he noticed that a mould had grown on the back of a photograph of his wife, Helen, which he had taken with him. He cultured the mould, which was a strain of *Penicillium funiculosum*, and tested the filtrate for its antiviral effects. It proved to contain an active principle which he named helenine, which inhibited the growth of many viruses without apparently being toxic. He published the results of his work [23], but it attracted relatively little attention, perhaps because most of those who tried to repeat his findings had no success. Nevertheless, scientists at Merck Sharpe and Dohme obtained a culture of the mould directly from Shope and confirmed that culture filtrates were indeed active. In further studies, they showed that the activity in helenine resulted from chronic infection of the *Penicillium funiculosum* strain, which Shope had fortuitously cultured with a fungal virus which contained double-stranded RNA. When minute concentrations of this dsRNA were added to cell cultures or injected into mice, they led to the appearance of a broad spectrum antiviral protein, which of course was by then readily identified as an interferon [24] Other dsRNA of natural or semi-synthetic origin similarly proved to be powerful inducers of interferon, Another broad spectrum antiviral agent, statolon, discovered earlier in the laboratories of Eli Lilly, ultimately proved to owe its activity similarly to a dsRNA-containing fungal virus [25].

With the benefit of hindsight, how significant have the interferons been? First, they have proved important agents for medical use. Indeed, used alone, or now increasingly in combination with other agents, IFN-α preparations are the only effective treatment for chronic virus hepatitis B and C, two of the most prevalent and serious diseases affecting mankind, and they are routinely used in the treatment of certain leukaemias and other forms of cancer. IFN-β is currently the most effective treatment for the relapsing remitting form of multiple sclerosis. I think that in the future, interferons will increasingly be also used in many other medical conditions, but almost always in combination with other agents, for it seems improbable that in vivo, interferons ever act alone.

Interferons have had another and in some ways perhaps more important function. As the prototype cytokine, their development has immeasurably speeded up that of all the other cytokines identified since. "The Interferon" was the first hitherto undefined biological activity to be given a name, which was immediately memorable, and implied that the activity was due to a specific protein. Interferons were the first cytokines to be highly purified, and ultimately purified to homogeneity and a specific biological activity per mg protein that at first seemed almost incredible; to be sequenced, cloned, and produced by recombinant DNA technology, and also by mass culture of human cells; to be shown to bind to specific cell surface receptors; to trigger special signal transduction mechanisms, leading to the activation of specified genes and the production of the proteins which mediate their activities; to be used in clinical trials; to be licensed by various regulatory agencies for use in particular medical indications; and to become agents now routinely used in medical practice, etc. All these developments have taken some 30 years. In contrast, the other cytokines, the interleukins, TNF, the colony stimulating factors, erythropoietin etc. have benefited from their example, and their development has been telescoped into a few years.

The unfortunate media-inspired "hype" of the 1980 era led many to think of interferons as the new panacea, and the reality was inevitably followed by a very negative reaction. Overnight, interferons vanished from news bulletins and the front pages of newspapers, and irritated doctors, who had been besieged by patients demanding interferon treatment, soon accused those making interferon preparations of chicanery or outright fraud. Today, even those who might be expected to know better often seem ignorant of what has actually been achieved by using interferons in medical practice. Indeed the eminent medical historian, Roy Porter [26], in a widely acclaimed recent survey of the major advances in medical science from the earliest times, unwisely continued his story to modern times. His comments about the value of interferons in cancer are both scathing and totally erroneous, and he makes no mention of their much more important use in infectious disease. It is unwise to anticipate the verdict of history, when a too early assessment may thus prove so disastrously wrong. Nevertheless, I personally have no doubt that when future historians review the advances in medical science in the twentieth century, they will rank the cytokines among the most important of these, and the discovery of the interferons as the most important single event.

References

1. Henco K, Brosius J, Fujisawa A et al (1985) Structural relationship of human interferon alpha genes and pseudogenes. J Mol Biol 185:227–260
2. Roberts, MR, Liu L, Guo Q et al (1998) The evolution of the Type I interferons. J Interferon Cytokine Res 18:805–816
3. Diaz M, Bohlander S, Allen G (1996) Nomenclature of the human interferon genes. J Interferon Cytokine Res 16:179–180
4. Allen G, Diaz MO (1996) Nomenclature of the human interferon proteins. J Interferon Cytokine Res 16:181–184
5. Foster G, Rodriguez O, Ghouze F et al (1996) Different relative activities of human cell derived interferon-α subtypes: IFN-α8 has a very high antiviral potency. J Interferon Cytokine Res 16:1024–1033
6. Pfeffer LM, Dinarello CA, Herbermann RB (1998) Biological properties of the recombinant alpha-interferons: 40th anniversary of the discovery of interferons. Cancer Res 58:2489–2499
7. Hibbert L, Foster G (1999) Primary human B cells proliferate in response to much lower concentrations of IFN-α8 than of other Type I human interferons. J Interferon Cytokine Res 19:309–318
8. Hoskins M (1935) A protective effect of neurotropic against viscerotropic yellow fever virus in Macacus rhesus. Am J Trop Med Hyg 15:675–680
9. Magrassi F (1935) Studi sul'infezione e sul'immunita da virus erpetico. III. Rapporti tra infezione e superinfezione di fronte ai processi immunitari: sulla possibilita di profondamente modificare il decorso e gli esiti del processo infettivo gia in atto. Z Hyg Infektionskr 17:573–620
10. Henle W, Henle G (1943) Interference of inactive virus with the propagation of the virus of influenza. Science 98:87–89
11. Henle W, Henle G (1944) Interference between inactive and active viruses of influenza. 1. The incidental occurrence and artificial induction of the phenomenon. Am J Med Sci 207:705–717
12. Henle W (1950) Interference phenomena between animal viruses: a review. J Immunol 64:203–236

13. Lennette EH, Koprowski H (1945) Interference between viruses in tissue culture. J Exp Med 83:195–219
14. Burnet FM, Fraser KB (1952) Studies on recombination with influenza viruses in the chick embryo. I. Invasion of the chick embryo by influenza viruses. Aust J Exp Biol Med Sci 30:447–458
15. Pieters T (1997) Shaping a new biological factor, '"the interferon", in room 215 of the National Institute for Medical Research, 1956/57. Phil Sci 28:27–73
16. Isaacs A, Lindenmann J (1957) Virus interference. I. The interferon. Proc R Soc Ser B 147:258–267
17. Fulton F, Armitage P (1951) Surviving tissue suspensions for influenza virus titration. J Hyg Camb 49: 247–262
18. Henle W, Henle G (1984) The road to interferon: interference by inactivated influenza virus. In: Billiau A (ed) Interferon, vol 1: general and applied aspects. Elsevier, Amersterdam, pp 3–22
19. Nagano Y, Kojima Y, Sawai Y (1954) Immunité et interférence dans la vaccine. Inhibition de l'infection dermique par le virus inactive. C R Soc Biol 148:750
20. Nagano Y (1981) Studies on virus-inhibiting factor or interferon. Kitasato Arch Exp Med 54:1–15
21. Ho M, Enders JF (1959) An inhibitor of viral activity appearing in infected cell cultures. Proc Natl Acad Sci USA 45:385–389
22. Ho M, Enders JF (1959) Further studies on an inhibitor of viral activity appearing in infected cell cultures and its role in chronic viral infection. Virology 9:446–477
23. Shope RE (1953) An antiviral substance from Penicillium funiculosum. II. Effect of helenine upon infection in mice with Semliki-forest virus. J Exp Med 97:627
24. Lampson GP, Tytell AA, Field AK et al (1967) Inducers of interferon and host resistance. I. Double-stranded RNA from extracts of Penicillium funiculosum. Proc Natl Acad Sci USA 58:782–789
25. Kleinschmidt WJ, Murphy EB (1967) Interferon induction with statolon in the intact animal. Bacteriol Rev 31:132
26. Porter R (1997) The greatest benefit to mankind. Harper-Collins, London

2 What Constitutes Therapeutic Success? The Interferons (1978–1998)

T. Pieters

2.1 Introduction

Every so often, during the past 25 years, scientific claims of therapeutic accomplishment in the treatment of cancer have entered the public domain. The biographies of these experimental therapies seem to have a familiar pattern in common: implied claims for a cancer cure based on preliminary tests, exaggerated coverage in the media, high public expectations, widespread disappointment and loss of interest when the claims failed to materialize in large scale trials, researchers and administrators forced on the defensive. and finally there seems hardly any reason for continuation of research or for clinical application.

A similar sequence of events unrolled in the course of 1978 following public statements by scientists who claimed to have achieved promising results with the biological substance interferon (IFN) in preliminary trials on cancer patients. As might be anticipated, the media gave enthusiastic coverage to the implied claims for a cancer cure which laid the foundation for a global interferon hype. The euphoria surrounding IFN as a "miracle cure" for cancer was short lived and faded when it appeared that both the performance of IFN in large scale cancer trials had been disappointing and that it often produced side-effects in patients (for extensive accounts of the early development of interferon see Pieters[1,2]).

However similar the course of events in the case of IFN, the most common dead-end scenario of potential cancer remedies did not materialize. Despite having failed to live up to its public promise as a therapeutic breakthrough, IFN succeeded in finding a niche in clinical practice. Today, thousands of patients suffering from multiple sclerosis, certain types of cancer and a number of virus diseases are routinely treated with IFN injections. Given the intense disappointment in the early 1980s in the healing power of what now became dubbed "the miracle drug looking for a disease", the following question deserves scrutiny: how did IFN manage to become legitimized as part of medical practice in the 1990s?

2.2   Beyond Interferon

As might be expected, the people working on IFN tried as hard as they could to account for the disappointments in order to safeguard funding. IFN researchers conveyed the impression that with more questions than answers they were just beginning to explore the potential of IFN. The diversity of IFNs with distinct and complementary activities seemed to grow every day, although clinical testing of the first IFN preparations produced with recombinant DNA technology had yet to start. The consensus view was, that although IFN as a single agent might turn out to be medically useful in treating viral infections, it might ultimately prove most valuable as part of the increasingly popular "multi-modality" approach in cancer treatment. IFNs could then be used as biological enhancers helping to increasing the host's own response against the tumor in combination with the three main cancer therapies of surgery, chemotherapy, and radiation [3–6].

By creating an image of IFN as a prototype of a promising, new, but still poorly understood area of cancer therapy known as immunotherapy, one that was going to play a significant role in future cancer practices, IFN promoters succeeded in establishing a more permanent base for support. The overall message was, as a science reporter of the *Washington Post* aptly expressed it in his headline, "Beyond Interferon" [7]. Cancer treatment centers that aspired to maintain an image of being at the cutting edge of the field of clinical oncology could not afford *not* to study an experimental therapy that was closely linked with the latest developments in tumor biology and molecular biology (E. Borden 1992, personal communication). Only the relative scarcity and impurity of IFN preparations withheld them from pursuing IFN research more actively. They waited for the more pure and homogeneous rDNA-produced interferons to become more widely available.

The apparently unremitting optimism within scientific quarters contrasted with the growing scepticism about IFN's potential among senior executives in pharmaceutical companies. Of the 20 or more firms which had announced their intention of investing in interferon development back in 1979, a growing number ceased work on IFN over a period of 1 year, namely during 1983. They had either lost out in the "cloning race" or management had negatively assessed IFN's potential both as a therapeutic drug and as a means to attract capital investment from the stock markets [8]. Apart from the disappointing cancer trial data there were also growing doubts about IFNs potential as an antiviral agent. The expectations, fuelled by reports from the Common Cold Unit at Salisbury that nasal sprays of alpha IFNs markedly reduced the occurrence of colds in volunteers, about opening up the commercially interesting niche market of common cold prevention were dashed with the latest news about its side-effects [9,10]. Even at low-dosage levels IFN appeared to induce side-effects similar to the symptoms responsible for most of the miseries of the very same cold that it was supposed to prevent. "A single high dose of zinc may do just as good a job at substantially lower cost", was the scathing judgement of the science correspondent, Joseph Alper, in *The New York Times* [11].

In the boardrooms of the three "IFN champions" that had most heavily invested in the substance, the Burroughs Wellcome Company ('Wellcome') and, most notably Hoffmann-La Roche ('Roche') and Schering-Plough, IFNs lacklustre clinical performance in treating tumors as well as virus infections was reason for growing concern [8; L. Gauci 1990, personal communication]. In 1983 Hoffmann-La Roche and Schering-Plough allocated 15% of their research budgets, more than $40 million each, to interferon. The company officials responsible for investment in IFN had a hard time defending the costly IFN research and development (R&D) programs. However, they succeeded in convincing the sceptics in the managements that although IFN might not be a "magic bullet" in itself, it had enormous potential as a prototype of a new generation of custom-designed biosynthetic drugs that was expected to generate a therapeutic revolution. They presented it in terms of the unique possibility of learning to tap an entire apothecary of new drugs from natural substances within the human body.

Unlike most chemical drugs these genetically engineered biologicals could be viewed in their "natural" (physiological) role as regulators or modifiers of a variety of pharmacologically interesting biological mechanisms. They argued that in order to develop these biosynthetic molecules as therapeutics, new testing methods and procedures would be needed to effectively bring them to the clinic. For instance, since human interferons were only active in humans and monkeys, researchers were forced to do the unusual and skip most of the preclinical animal testing and go straight into humans to determine a toxicological and pharmacological profile.

However different, work on biologicals at the same time was argued to be easily integrated into the current system for drug development based on tinkering with molecules as a means to create variants with most optimal pharmacological characteristics in terms of risk: benefit ratios. The promise of scientific and therapeutic innovation appealed most to senior drug company executives who were less worried about today's profits than about tomorrow's prospects. With patents on most of the top prescription pills expiring by 1990 and with unprecedented few potential "blockbuster" products in the pipeline, drug companies had come to recognize that a new wave of innovation was needed to position themselves for long-term survival [12; L. Gauci 1988 and J. Warbee 1988, personal communications].

However, those responsible for the IFN-related R&D programs were not able to get round the iron business rule of profit generation. The pressure for return on investments in IFN research and in production facilities for the first genetically engineered IFN products (Schering-Plough's homogeneous interferon alpha-2 and differing by only one single amino-acid, Roche's homogeneous interferon alpha-A preparation) and Wellcome's "natural" IFN product, containing the full spectrum of alpha IFNs produced by cultured human cells, did force the IFN advocates within the companies to adapt their R&D strategies. In line with the government supported research programs, the pharmaceutical industry focused on IFN as a prototype of a new generation of tailor-made recombinant biologicals to be used as therapeutics in the new kind of disease management: immunotherapy within a multi-modality treatment framework. Apparently the drug companies recognized the strategical and commercial importance of taking advantage of the more general move across medicine towards combination therapy (J. Petriccianni 1992, personal communication).

However, the drug regulatory authorities found the combined-modality approach difficult to assess as their evaluative practice and standards were still governed by a single-agent therapeutic philosophy. For interferon to be considered legally as a new therapeutic drug, it had to be officially evaluated as a single agent. This implied that before licensing procedures could be taken into consideration, the companies had to look for a disease, rare though it might be, that justified a need for interferon (J. Petriccianni and L. Gauci, personal communications, for similar cases of drugs looking for diseases see Vos [13] and Oudshoorn [14]). With most trial responses to single-agent IFN therapy comparing unfavourably with drugs already available in the clinic, the industrial IFN program managers faced the seemingly Herculean task of establishing an unambiguous justification for clinical use.

2.3 Naturalizing Interferons as Cytokines

In search for suitable diseases as candidates for interferon as a treatment, the drug companies actively supported clinical trials to evaluate the effects of IFN on as wide a variety of diseases as possible. They offered clinical investigators world-wide   large quantities of their IFN products free of charge. IFNs were tested against hepatitis B, various lymphomas, colds, breast cancer, prostatic cancer, multiple sclerosis, herpes keratitis, malaria, AIDS (as soon as the epidemic began to manifest itself around 1982) and many other cancer and virus-related disease conditions. The drug companies mounted one of the most intensive clinical trial programs ever set up to evaluate a new pharmaceutical agent. In the first 3 years (1982–1984) the recombinant IFNs alone were tested in over 4000 patients [15–18].

By the end of 1983 the large-scale testing efforts finally seemed to show signs of paying off. Drug company officials swung into action as soon as the news broke in the fall of 1983 about Jordan Gutterman's research group of the M.D Anderson Hospital and Tumor Institute in Houston (USA) having achieved a higher than 80% trial response to "natural" alpha IFN therapy in patients with a rare form of chronic leukaemia known as "hairy-cell leukaemia" [19].

The knowledge that there was hardly any viable treatment for this form of cancer with estimated mortality rates of about 15% per year, sufficed for those looking for a suitable disease indication [20; L. Gauci and J. Petriccianni, personal communications]. They seemed to take for granted the fact that among clinical oncologists scepticism prevailed about the overwhelming efficacy claim by Gutterman, who had earned himself a name for overoptimistic interpretation of IFN trial data [8]. The first priority of companies was to demonstrate that the dramatic clinical effect could be reproduced by other clinical researchers. Following extensive clinical testing they ultimately succeeded in their efforts. Consequently licence applications for what Roche and Schering-Plough both dubbed "the world's first anticancer interferon" were sent for evaluation to, among other drug regulatory agencies, the FDA [21]; in contrast to interferon products which were still in the investigational stage and limited in their clinical use (phase I,II,III testing), licensed products which would be available for general commercial distribution and use by the medical profession must meet a variety of extra regulatory requirements, at least in the United States, as prescribed by the Good Manufacturing Practice (GMP) provisions and the General Provisions for Licensed Biologicals (GPLB) Code of Federal Regulations [22].

It is interesting to see that Wellcome with its non-recombinant "natural" IFN product was forced to follow a slightly different course. In most western countries like the USA and the Netherlands, Wellcome's natural alpha IFN "cocktail" (with the brand name Wellferon) met with difficulties from the regulatory authorities except for Britain, the homemarket for Wellcome products. Asiatic countries, which historically do not oppose heterogeneous pharmaceutical preparations because of a different therapeutic drug culture, had no problem accepting the IFN cocktail. By the time the FDA and Dutch drug regulatory officials put aside their doubts about Wellferon, Wellcome decided that with the market already flooded with Roche's and Schering-Plough's-rDNA produced IFNs, it was no longer commercially interesting to apply for a product license (interview with a Dutch Glaxo-Wellcome worker who prefers not to be identified; Kari Cantell's Finnish leucocyte interferon 'cocktail' met with the same difficulties from the drug regularity authorities [23]).

In the meantime the therapeutic development of IFN had helped to spawn a spectacular investigative enterprise. This included numerous academic, federal and industrial research programs, the *Journal of Interferon Research* and the formation of the International Society of Interferon Research. Going by the growth in the number of papers published each year with the word "interferon" in their titles from about 800 in 1980 to more than 3,000 in the year 1987, the field experienced an information explosion (, for the annual number and titles of interferon-related publications see the Dindi data base, Cologne).

The fact that the accumulation of laboratory data had already reached impressive proportions without much progress in the understanding of the mechanisms of action was not regarded as a failure but instead was used to advertise IFNs unwieldy and complex nature as part of an intricate system of checks and balances. This so-called cytokine network was said to constitute a key part of the body's natural defenses. The complexity of the cytokine interactions was argued to complicate IFNs study in the test tube as well as its clinical application. Researchers maintained that additional knowledge of these interactions was needed to improve or optimize the use of cytokines like IFN. The questions piled up but there were few satisfactory answers. Did all IFNs have the same or different functions within this network? Which IFNs should be tested first clinically, on what diseases and how? Was it justified to start from maximum tolerated dosage regimens knowing that more did not always appear to be better with these substances? Would combination treatment indeed add up to more than the sum of its parts? IFN advocates called it difficult substances to work with, far more complex than traditional therapeutic drugs. As a justification for their research efforts they kept emphasizing that IFN generated intriguing research horizons. It was widely claimed to provide a window both into biological processes at the molecular level and new approaches to the treatment of disease [24–28].

Laboratory and clinical researchers alike stressed that the clinical potential of interferons would be enhanced by further investigation and understanding of the molecular mechanisms underlying the effects of interferon in patients. The solidarity implicit in both their emphasis on the importance of linking biomedical research at the cellular and molecular levels to clinical interventions as part of what became known as "molecular medicine", stemmed from shared interests in the maintenance of credibility and funding (for a focus on what constitutes today's molecular medicine see the web site http://www.isinet.com/prodserv/focus_on/fo_mmfrm.html). Disappointing as well as successful clinical results could be used by bench scientists to claim that further basic research was mandated. At the same time clinical researchers were able to defend their poor results by referring to the slow progress in the understanding of the underlying mechanisms by which interferons exerted their biological effects both in vitro and in vivo. Achieving occasionally high response rates at the bedside, however rare the disease condition, seemed to justify the further expansion of both preclinical and clinical research. Nonetheless the actual expansion of interferon related activities was largely dependent on the support and coordination of the pharmaceutical industry.

During the transition to the large-scale production of genetically engineered IFNs, laboratory scientists and clinicians had become increasingly dependent on the industrial infrastructure. In possessing the key to the material resources and research materials the companies began to play a central role in structuring the development of IFN (L. Gauci, personal communication; [29–31]). Laboratory scientists were encouraged to deploy the new techniques of molecular biology, cellular physiology and biochemistry in the search for additional naturally occurring molecules that might provide genetic blueprints for the development of these future therapeutic drugs [32]. Moreover, by providing a new powerful research tool, unlimited quantities of pure crystalline IFN, the research community was stimulated in efforts to investigate the underlying molecular mechanisms of IFNs actions [33]. At the same time clinical investigators were encouraged to start experimenting in humans with this first representative of a rapidly growing family of clinically active biologicals, so-called "cytokines". The impression was conveyed that participating clinicians would stand out as pioneers of a new era of disease treatment ('Interferon in prospect', a three-part film series from Schering-Plough Corporation, USA, that was made available in 1983 to clinical investigators all over the world).

In 1986, once the FDA, swiftly followed by other regulatory agencies, licensed Roche's and Schering-Plough's recombinant alpha IFN products for the treatment of hairy cell leukaemia, the marketing branches of the companies were fully activated as a means to establishing a growing need for IFN (T. Pike 1990, personal communication). In order to interest health care professionals in the new area of IFN therapy and to convince them of its scientific character, Hoffmann-La Roche and Schering-Plough established freely accessible electronic data bases (e.g. Schering-Plough's *ICON*), specific journals (e.g. Roche's *Progress in Oncology, Progress in Virology* and *Interferons Today and Tomorrow*), and financed medical book publications (e.g. Schering-Plough sponsored the publication of *Interferons in Cancer Treatment*); [34–37]; ICON or Interferon Communication Network was established in 1980 by Schering-Plough. In 1988 ICON contained about 15,000 items related to interferon which had been selected by a group of 5 physicians payed for by Schering-Plough. The Schering-Plough team produced their own abstracts of the papers and on special request customers could obtain the original publications (C. v. Helsen 1988, personal communication).

Optimising response rates seemed to be the explicit aim of virtually any clinical research project dealing with IFN [38–40]. Except for hairy cell leukaemia clinical researchers involved in testing IFNs claimed response rates anywhere between 10% and 50%. The problem and advantage of talking in terms of percentages was that success appeared to be a highly ambiguous term. Overall response rates ("efficacy") resulting from clinical studies under controlled circumstances might look promising, even when it remained unclear what this actually meant in terms for individual chances of success and for how well a treatment might perform in everyday clinical practice ("effectiveness"; my distinction between effectiveness and efficacy is derived from clinical epidemiology; see Pocock [41]). Regardless of interpretation, however, response rates remained invariably low in most diseases, suggesting that it could help only some of the patients some of the time.

Under normal disease conditions this kind of negative scientific assessment would dissuade doctors from applying a therapeutic drug. But in circumstances where there is no hope for a cure the rules of the game are different for both doctors and patients. In diseases in which success treatment is rare, seeking treatment through medical intervention is a gamble which by definition can have few winners. Gambling is an alluring analogy for all parties as it turns poor clinical results into acceptable chances, allotting responsibility for failure to bad luck rather than medical or other capacities (for a similar kind of phenomenon in the treatment of reproductive disorders in women see [42]).

In 1986 the Dutch IFN researcher Huub Schellekens predicted that IFN would be used more widely in 10 years time than might be expected on the basis of the approved list of indications because of the psychology of IFN as a potential "multi-drug" from common colds to cancer (for interview with Schellekens H, 'Intron A' see press information video tape produced by Schering-Plough in 1986 on the occasion of the market introduction of their interferon product). Schering-Plough and Roche of course fully endorsed this perceived mechanism and were keen to turn this into a self-fulfilling prophesy. They hoped that if they were able to show that IFN as a cytokine enhanced – even if marginally – the efficacy of routine and semiroutine therapies for problematic cancers and chronic virus diseases, the molecule could become a commercial success as an auxiliary immunotherapeutic drug. They therefore actively promoted trials and investigator's meetings that would evaluate the possible synergism between IFNs and various conventional therapies. At the same time they began marketing IFN as part of the comprehensible though open-ended (in terms of number of factors involved and nature of interactions) cytokine network. In the process the Roche sponsored journal, *Interferons Today and Tomorrow*, changed its name into *Interferons and Cytokines*, while *The International Society for Interferon Research* became *The International Society for Interferon and Cytokine Research*, and the official journal of this society, became the *Journal of Interferon and Cytokine Research* by 1995 (according to an editorial by Ronald Penny the new journal title reflected the real breadth of focus in this ever expanding field. In his view the number of factors involved grew "almost at the rate of action of the cytokines themselves" [43]).

With the relentless support of the drug industry and patients in desperate need for a cure, and through the combination of scientific drive and professional ambition, most notably clinical oncologists and infectious disease specialists continued to tinker with the design of trials. They tried different combinations (e.g. in combination with cytotoxics or a combination of various biological response modifiers) and different routes and durations of administration [39,44]. In doing so they ultimately tinkered toward success in terms of establishing new therapeutic drug practices for IFN and actively working on the treatment's effectiveness. Although superior treatments for the treatment of hairy cell leukaemia became available, and have largely replaced the use of IFN for this indication. IFN as an adjunct to other therapeutic modalities became part of the routine treatments of a growing number of major malignancies such as non-Hodgkin's lymphoma, multiple myeloma and kidney cancer.

IFN, almost given up on in terms of having a role to play in medicine by business analysts, the press, the public, scientists and doctors alike in the early 1980 s, has become exemplary not only for a new generation of biosynthetic drugs but also for the health benefits that result from advances in molecular medicine. Fuelled by a potent combination of scientific drive, professional ambition, marketing efforts and a lot of hard work between bench and bedside, IFN was transformed from an orphan-drug into a billion-dollar molecule; the world-wide market for IFNs expanded from $13.6 million in 1986, to $751 million in 1992, to $1.57 billion in 1993, down to $928 million in 1995 and up to an estimated $1.2 billion in 1998 according to the December 1997 issue of *Pharma Business Signals* (Competitive Intelligence and Strategic Issues/Trends).

The widespread application of interferon as an effective medical commodity seems almost beyond dispute; the need for it appears to be a premise, rather than a contestable assumption. As a consequence, opposition to interferon therapy currently revolves less around questions of need than around questions of cost or economic feasibility, which increasingly dominate the political agenda of "marketplace" medicine. As the chapter on economic aspects of interferon in the 1997 book *The Clinical Applications of the Interferons* shows, economic and commercial considerations, are easily linked with medical discourse and therefore are difficult to separate from medical arguments showing the necessity for interferon treatment [45].

2.4 Conclusions

The last paragraph on economics nicely illustrates that shaping therapeutic drug practices is not exclusively a medical or scientific affair. To think so would deny the multiple links between science, medicine and society which have been made manifest in my description of the development and dissemination of interferon as part of a new therapeutic approach in the treatment of cancer and viral disease. Of course, in order to become part of the therapeutic armament of doctors, interferon had to meet scientific standards of what constituted legitimate evidence for efficacy. But does meeting those standards suffice to explain the exponential growth in the clinical use of a variety of interferons?

As far as the spokespersons of the pharmaceutical companies are concerned it does. The growing market for interferon treatment is claimed to be a "natural consequence" of a growing need of doctors and patients ("consumers") who apparently regard the new drug as a valuable asset. The growing range of therapeutic effects, so they argue, reflects the need for interferon. However, if there is one thing the interferon crusade can teach us, it is that the equation of need and the widening of the pharmacological spectrum, is a logical fallacy. The need for interferon was never "simply" there. Likewise, the development and dissemination of interferon, was never a self-evident, inevitable process. This brings us back to the question how did interferon manage to become part of the doctor's bag?

First of all, confronted with the erosion of public support for the interferon crusade, attempts were made to develop alternative strategies to legitimate work on interferon. Instead of the presumed non-toxic nature of interferon, its perceived unique capacity as a biological to act through the immune system was used to attract attention. Moreover, interferon was presented as having an important advantage over conventional therapies: it linked the clinics to advanced laboratory research in tumor biology, molecular biology and immunology. I dubbed the rhetorical strategy that proved effective in establishing a more permanent base for support "beyond interferon": picturing interferon as a first step in the development of immunotherapy, a promising new therapeutic approach in the treatment of cancer, that was going to constitute a medicine of tomorrow.

Second, the panacea image that was attached to interferon continued to appeal equally to defenders of medical orthodoxy and advocates of unorthodox cancer remedies. However alien to a medical practice that was dominated by the theory of specific etiology and the notion of specific therapy, interferon had an irresistible appeal to laboratory researchers and clinicians alike. It was the promise of a new flexible immunological approach to disease which might make it possible in the future to individualize disease treatment through forging further links between the bench and the bedside.

Third, by 1980 strong links had been formed between interferon research, the booming field of molecular biology and private industry. Powerful commercial, institutional and professional interests had become aligned with what was considered a major show case of the newly developing biotechnology industry and its revolutionary rDNA technology. Originally deployed only with the aim to make highly purified interferon preparations available in large quantities, at modest cost, the genetic engineers, closely linked with the pharmaceutical industry, played a major role in redrawing the interferon landscape.

The genetic engineers opened the door to an alternative approach to drug development. They promised to create a new horizon in pharmaceutical research by engineering an entire apothecary of new biosynthetic drugs from natural substances within the body. As promoters of this new approach, they stressed novelty but at the same time they presented genetic engineering as fitting in well with the classic chemistry-based therapeutic drug development practice. The promise of a new approach to drug development kept the pharmaceutical companies, who had begun investing on a large scale in interferon as a promising therapeutic lead and demonstration project of rDNA production technology at the end of the 1970s, interested when the news about its less-than-spectacular performance in clinical trials began to spread in the course of 1980.

During the subsequent successful transition to large-scale production of interferon, the international pharmaceutical industry, which was seeking to position itself for long-term survival through scientific and therapeutic innovation, became the most dynamic and strategic actor of all those involved in work on interferon. In achieving a key position in the distribution of research resources and materials, the pharmaceutical industry increasingly dictated the development of interferon. However dominant their role in structuring interferon's biography, for the profiling and marketing of interferon the drug companies were largely dependent on their ability to enter into cross-fertilizing relationships with laboratory scientists, clinicians and regulatory bodies.

I have shown that in order to become accepted, interferon needed a therapeutic profile and treatment concept that could be integrated or combined with existing therapeutic practices. This required that the drug makers in collaboration with laboratory researchers and clinicians actively created and made available a sort of rough-and-ready "therapeutic rationale". Clinicians and laboratory scientists alike increasingly perceived cancer as a family of diseases in which complex cellular and molecular regulatory functions have gone awry, instead of a unitary disease characterized by the uncontrolled growth of malignant cells that spread out aggressively and destroy the body.

Talking about the "Big C" as a multifactorial molecular disease was consistent with the emergence of immunotherapy as part of a combined-modality, laboratory-supported therapeutic framework in clinical oncology. This linked up perfectly with the efforts in the field of interferon research to redefine interferon as a means to develop a new complementary modality of cancer treatment that might be used in conjunction with conventional therapies. Redefining interferons as cytokines which could be used therapeutically to strengthen the body's self-defense as part of a laboratory-guided multi-modality therapy framework, legitimated and reinforced the new molecular perception and treatable nature of cancer. And this allowed all parties to see interferon therapy as therapeutic. To put it differently, as advertised products of molecular biology the interferons fitted in well with the "molecularisation" of medicine and this helped to make them "work". I pointed to the solidarity implicit in the emphasis of both laboratory researchers and clinicians on the importance of linking through interferon bench and bedside as part of a new kind of molecular medicine. In always being able to justify that either further laboratory or clinical research was mandated, it made both groups relatively immune to disappointing research news.

As a final point, the "industry" of clinical trials not only proved effective in widening IFNs therapeutic profile but also in marketing the combined modality therapeutic concept that turned the IFNs from unwanted drugs into top-selling pharmaceuticals. By picking up and refining the broadly based cytokine network concept in such a way that it served the needs of all those involved, the drug makers anticipated the successful integration of interferon into medical practice. In positioning interferon as a helpful *neighbor* which was compatible with and supportive of existing treatment practices, the pharmaceutical industry succeeded in having interferon absorbed relatively quickly by the medical infrastructure.

References

1. Pieters T (1993) Interferon and its first clinical trial: looking behind the scenes. Med Hist 37:270–295
2. Pieters T (1997) History of the development of the interferons: from test-tube to patient. In: Stuart-Harris R, Penny R (eds) The clinical applications of interferons. Chapman and Hall, London, pp 1–19
3. Krown S (1981) Prospects for the treatment of cancer with interferon. In: Burchenal J, Oettgen H (eds) Cancer; achievements, challenges, and prospects for the 1980's. Grune and Stratton, New York, pp 367–379
4. Johnson R (1981) Interferon: cloudy but intriguing future. JAMA 245:109–116
5. Newmark P (1981) Interferon: decline and stall. Nature 291:105–106
6. Sun M (1981) Interferon: no magic bullet against cancer. Science 212:141–142
7. Fenyvesi C (1981) Beyond interferon. The Washington Post (14-6-1981), p 28
8. Powledge T (1984) Interferon on trial. Biotechnology 2:214–228
9. Editorial (1982) Million dollar cold cure. The Sunday Times (3-10-1982), pp 22–30
10. Scott GM, Phillpotts RJ, Wallace J et al (1982) Prevention of rhinovirus colds by human interferon alpha-2 from Escherichia Coli. Lancet ii:186–187
11. Alper J (1994) First there was interferon. The New York Times (18-11-1994), F 13
12. Wycke A (1987) Molecules and markets. The Economist (7-2-1987)
13. Vos R (1991) Drugs looking for diseases. Innovative drug research and the development of the beta blockers and the calcium antagonists. Kluwer Academic, Amsterdam
14. Oudshoorn N (1994) Beyond the natural body: archeology of sex hormones. Routledge, New York
15. Editorial (1983) Interferon may help AIDS victims. New Scientist (3-11-1983)

16. Editorial (1981) Interferon tested on sclerosis. The New York Times (21-11-1981)
17. Finter N, Oldham RK (eds) (1985) Interferon, vol 4: in vivo and clinical studies. Elsevier, Amsterdam; (11-2-1983) update on interferon. News Service American Cancer Society, NCI Archives File no DC8301–006691
18. 'Intron A', press information video tape produced by Schering-Plough in 1986 on the occasion of the market introduction of their interferon product
19. Quesada J, Hersh E, Gutterman J (1983) Hairy cell leukemia: induction of remission with alpha interferon. Blood 62:207a
20. Quesada J, Reuben J, Manning J, Hersh E, Gutterman J (1984) Alpha interferon for induction of remission in hairy cell leukemia. N Engl J Med 310:15–18
21. Editorial (1983) Anticancer interferon available soon. Hospital Doctor (22-9-1983)
22. Petricciani J, Esber E, Hopps H, Attallah A (1984) Manufacture and safety of interferons in clinical research. In: Came P, Carter W (eds) Interferons and their applications. Springer, Berlin Heidelberg New York, pp 357–370
23. Hage H (1986) Controlerende instanties moeten soepeler worden met interferon. Toegepaste Wetenschap TNO (1-10-1986)
24. Baron S, Dianzani F, Stanton G, Fleischmann W (eds) (1987) The Interferon system: a current review to 1987. The University of Texas Press, Austin
25. Baron S, Coppenhaver DH, Dianzane F (1992) Interferon: principles and medical applications. University of Texas Press, Austin
26. Wallis C (1985) What's become of interferon? Time (1-6-1985)
27. Balkwill F (1985) Interferons: from common colds to cancer. New Scientist 105:26–28
28. Balkwill F (1989) Cytokines in cancer therapy. Oxford University Press, Oxford
29. Gauci L (1990) Interferon drug development: a history truely consistent with the discovery process, paper presentation at the international meeting 'from clone to clinic', Amsterdam
30. Oldham RK (1985) Interferon: a model. In: Gresser I (ed) Interferon 6. Academic, London, pp 127–141
31. Barnes D (1987) Biologics gain influence in expanding NCI Program. Science 237:848–850
32. Foerstner A (1985) How we make our own wonder drugs. Chicago Tribune (20-1-1985)
33. Johnson HM, Bazer FW, Szente BE et al (1994) How interferons fight disease. Sci Am 270:40–47
34. Thomas HC, Cavalli F, Talpaz M (eds) (1987) Thirty years of interferon. Mediscript, London (Interferons today and tomorrow,vol 5)

35. Kirchner H (ed) (1986) Update on interferons. Progress in oncology, vol 2. Mediscript, London
36. Kirchner H (ed) (1987) Update on interferons. Progress in virology, vol 1. Mediscript, London
37. Silver HK (ed) (1986) Interferons in cancer treatment. Medical Education Services, Mississauga , Canada
38. Came PE, Carter WA (eds) (1984) Interferons and their applications. Springer, Berlin Heidelberg New York
39. Pinsky C (ed) (1986) Biological response modifiers. Semin Oncol 13:131–227
40. Parkinson D (ed) (1994) The expanding role of interferon-alfa in the treatment of cancer. Semin Oncol 21:1–37
41. Pocock S (1983) Clinical trials. Wiley, Chichester
42. Van Dyck J (1995) Manufacturing babies and public consent. New York University Press, New York, p 124
43. Penny R (1990) Editorial. Interferons Cytokines 15:3
44. Stuart-Harris R, Penny R (eds) (1997) Clinical applications of the interferons. Chapman and Hall, London
45. Shiell A, Salkeld G (1997) The economic aspects of interferon. In: Stuart-Harris R, Penny R (eds) The clinical applications of interferons. Chapman and Hall, London, pp 376–390

3 The Prehistory and History of the Uses of Interleukin-2 in Cancer Therapy *

I. Löwy

3.1 Prelude: The Multilayered History of Cytokines as Antitumor Drugs

Numerous anticancer therapies (radiotherapy, nitrogen mustard, anti-folic acid antagonists) have been developed following accidental clinical observations. The antitumor properties of other compounds have been revealed through systematic screening programs, such as the natural substances screening program of the National Cancer Institute. The

---

* My book "Between Bench and Bedside" [1] discusses in more detail the introduction of interleukin-2 into cancer therapy

history of these therapies is relatively short, and it includes original clinical or biological observations, usually (but not necessarily) tests in laboratory animals, and finally trials in patients. The story of the introduction of interleukin-2 into cancer therapy is different. It is linked with the history of immunology of cancer, an enterprise which started more than a 100 years ago and has been rich in abandoned trails. The introduction of interleukin-2 to cancer therapy was the result of a partly contingent encounter between several distinct lines of investigation. My paper follows the complex history of the attempts to develop immunotherapies of cancer, and stresses the intertwining – and the separation – of different strands of inquiry. It illustrates the early failures to transform promising observations about the putative role of immune mechanisms in the control of malignant growths into accepted therapies of human cancer. These early attempts are contrasted with the successful development of recombinant protein drugs in the 1980s. The increased scale and scope of biomedical investigations, and the consolidation of links between laboratory studies, bedside observations, large-scale clinical trials and collaboration with the industry, led to the transformation of a substance first described as an agent which promotes the growth of immunocompetent cells in the test tube into an anticancer drug.

3.2 Cells, Immunity and Cancer

The antitumor activity of interleukin-2 was attributed to its capacity to activate "killer lymphocytes" which eliminate malignant cells. Accordingly, histories of interleukin-2 often start with the statement that in the 1970s researchers confirmed that certain sub-classes of white blood cells – the so called cytotoxic (that is, cell-killing) lymphocytes – are able to destroy tumor cells in the test tube, and seem to play a role in the destruction of malignant growths in tumor-bearing laboratory animals. Interleukins, it was (and is) believed, stimulated the sub-populations of lymphocytes with putative antitumor activity [2,3].

The description of "cytotoxic lymphocytes" is usually presented as a direct consequence of studies of the cellular mechanisms of immunity in the 1960s and 1970s. This presentation is not correct. The idea that lymphocytes are involved in "resistance" to malignant tumors was proposed, then successfully tested in the laboratory in the early twentieth century. This idea was abandoned, however, and rediscovered 50 years later. In the nineteenth century the definition of cancer as an abnormally-growing tissue was based on anatomical and pathological observations, and cancer researchers at first explained malignancy as resulting from prolonged irritation or inflammation of the tissues [4]. The discovery that infectious diseases are induced by specific microorganisms triggered the search for a putative "cancer germ(s)". This search, together with the observation that tumors grafted into laboratory animals were usually rejected by the body's "defense mechanisms", led researchers to connect cancer studies and "resistance" to infectious diseases. (The search for a "cancer bacillus" was abandoned in the early twentieth century. In contrast, the debate on the role of viruses in the origins of human cancer which started in 1911 with the description of transmissible tumors of the fowl described by Rous [5], continues today too. Cf. Gaudillière [6].) Early studies of immunity were interested in both humoral and cellular mechanisms of immunity. They were divided into two chapters of near equal importance: studies of "natural immunity," that is, resistance to infectious disease without prior contact with the disease-inducing microorganisms (for example, such mechanisms explained why certain animal species are refractory to pathogenic germs that affect other animal species), and studies of "acquired immunity", subdivided into "natural acquired immunity", the consequence of an earlier encounter with the disease-inducing agent, and "artificial acquired immunity," resulting from vaccination. Cells (macrophages) were viewed as more active in "natural immunity", while humoral antibodies were more active in "acquired immunity." This distinction was, however, far from being absolute. Scientists who studied immune responses in the early twentieth century proposed that cells may become more specific during an immune response, while the serum contained both specific and non-specific antibacterial substances [7–9].

The broad perception of immunity as a physiological phenomenon which includes all the mechanisms of resistance to infection – specific and non-specific, humoral and cellular – disappeared however in the 1910s and 1920s. At that time the study of immune mechanisms was gradually identified with the investigation of "specific resistance" to infection, while the latter became synonymous with the study of specific antibodies that appear in serum following disease or vaccination. Studies of the "natural resistance to infectious disease" and of the cellular aspects to immunity were abandoned because of the technical and conceptual difficulties involved in this kind of research. The terms "natural immunity," and "resistance to infectious disease" covered a wide range of poorly defined pathophysiological phenomena, very difficult to standardize and to quantify with the existing immunological methods. By contrast, it was relatively easy to apply quantitative, reproducible methods for the study of antibodies. Specific antibodies in the serum, to put it in a nutshell, were a much more rewarding subject of study than elusive and obscure non-specific cellular "resistance mechanisms." The efforts of immunologists in the 1910s and 1920s were centered on the elaboration and calibration of quantitative methods of studying antibodies in the test-tube. These quantitative methods also had applied practical applications: serodiagnostic (identification of pathogenic microorganisms using specific antibodies), and serotherapy (the treatment of infectious diseases with specific antibodies [10]). These practices led to better integration of immunology laboratories in clinical settings, and to the birth of a new professional community: the serologists. During the so-called chemical period of immunology, serologists who studied antibodies, not pathologists who studied tissues and cells, became identified with the investigation of immune reactions [11].

The discovery of the role of lymphocytes in destroying tumors was made by Dr. James Bumgardner Murphy of the Rockefeller Institute, New York. Murphy joined the cancer laboratory of the Rockefeller Institute in order to assist its director, Dr. Peyton Rous, in the study of a "filterable agent" (supposedly a virus) able to induce malignant tumors in fowl [5]. Looking for ways to keep chicken tumors alive in the laboratory, Dr. Murphy developed a method for grafting tumors onto the embryonic membranes of fertilized eggs. He had noticed that, while tumor grafts in adult animals are always species-specific, it was possible to graft tumors onto embryos of different species. Not only was he able to graft fowl eggs with tumors from other birds, but he also managed to grow mouse and rat tumors in chicken embryos [12]. The grafted tumor thrived until the last 2 days of embryonic life. At that time a "resistance reaction" occurred which, Dr. Murphy noticed, was closely connected to the accumulation of lymphocytes around the graft. Such a "lymphocytic reaction" was not unique to the embryo. Dr. Murphy observed a similar reaction in adult animals that rejected previously established tumors. It is highly probable, he concluded, that the "lymphocytic reaction" was a physiological mechanism by which the body was able to get rid of unwanted cells and tissues [13,14].

If lymphocytes are indeed able to destroy malignant tumors in the body, Dr. Murphy reasoned, an increase in the amount and the activity of lymphocytes should improve the body's ability to fight malignant growth. In 1915 Dr. Murphy (together with Dr. John J. Morton) started a series of experiments aimed at demonstrating that a non-specific stimulation of the lymphocytes in a tumor-bearing animal can stimulate its resistance to tumors. They had shown that mice irradiated with low doses of X-rays, and which had increased numbers of lymphocytes in their blood, showed increased "resistance" to grafts of transplantable tumors but also to the re-implantation of their own, spontaneously generated tumors. The last results, Dr. Murphy believed, indicated that the non-specific stimulation of lymphocytes may lead to a cure for cancer in humans [15,16]. Dr. Murphy first announced his discovery in a paper read before the National Academy of Sciences (U.S.A.) on August 15, 1915. Dr. Simon Flexner, the Director of the Rockefeller Institute, viewed the observation as important, and made the paper's content public in a special press conference.

Murphy and Morton's study was greeted by the press as the announcement of a breakthrough in cancer studies. The promise, implicit in this investigation, that the new science of immunity would be applied to "vaccination" against cancer, clearly appealed to the public's imagination. Murphy and Morton's study was prominently reported in newspapers under headlines such as "John D. Doctors Believe Cancer Prevention Found," "Cancer Foe Discovered: Rockefeller Institute Scientists Find Immunity Guaranty", "Rockefeller Aides See Immunity From Cancer: Institute's Investigators Discover Means They Hope Will Make Man Proof Against the Disease", "Can Prevent Cancer Now: Rockefeller Institute Investigators Say by Increasing the Number of Lymphocytes the Cancerous Growth is Prevented" [17]. (On the attitude of the American press to cancer research at that time, see Patterson [18].) Over the course of the next 7 years Dr. Murphy and his collaborators were able to confirm and extend their original findings by showing a clear-cut correlation between the stimulation (or inhibition) of lymphocytes and the development of experimental tumors [19]. They were also able to start clinical studies rapidly.

In the clinical investigations conducted by Dr. Murphy, breast cancer patients were treated with low doses of X-ray radiation immediately after tumor removal, in the hope that the stimulation of their lymphocytes would enable the body to stop the progress of malignant disease. Dr. Murphy followed the evolution of the blood counts of patients who were given X-ray treatment, hoping to find a correlation between the amount of lymphocytes in the blood and the clinical progress of the disease. Unfortunately, no such correlation was found. The clinical experiment lasted from 1916 to 1922, with no noteworthy results. There was no improvement in patients' health and no relevant connection was found between clinical manifestations of cancer and patients' lymphocyte count. In his final observations on the clinical experiments Dr. Murphy noted that the data obtained in the human studies did not justify publication [20]. In 1920 Dr. Flexner no longer stated that the goal of Dr. Murphy's clinical studies was to find a cure for cancer. Much more modestly, Dr. Murphy's research was redefined as a study of the biological effect of X-rays on the human body, aiming at the elaboration of guiding principles for the use of radiation treatment in humans [21]. Dr. Murphy himself declared that tumors grafted in laboratory animals were imperfect models for the understanding of the spontaneously occurring human malignancies, and he turned to the study of chemically-induced tumors, for him a better model for the investigation of the malignant transformation of the cell. His suggestion that non-specific stimulation of the lymphocytes may play a role in finding a cure for cancer was gradually forgotten, and studies of the possible role of lymphocytes in the destruction of malignant tumors were neglected for the next 50 years.

In the 1920s and 1930s Dr. Murphy's criticism of the use of grafted tumors in cancer research was echoed by other cancer experts. Investigators who used grafted tumors, these critics proposed, were often unaware of the fact that tumors can be successfully grafted only on genetically identical animals. As a consequence, the rejection of grafted tumor was at first attributed to "resistance" to malignant cells, not to the rejection of foreign tissue. Gradually, however, scientists became aware of the intrinsic limits of studies made with grafted tumors in genetically heterogeneous animals [22]. In the 1920s,"resistance" to malignant tumors continued to be studied in numerous pathology laboratories, but these investigations lost much of their earlier prestige. A review of these studies, written in 1929, reflects this change [23]. Many investigators who pretended to study "resistance" to tumors, its author affirmed, did not pay attention to investigations that stressed the importance of working with a pure breed of mice (that is, mice that are genetically homogeneous), and the necessity to include in each experiment a large number of animals in order to overcome the problem of diversity in individual responses to malignant tumors. Moreover, the experiments made with animals, even those few which were correctly planned and executed, did not shed any new light on the problem of therapy or prevention for human cancer [24]. The whole field of application of immunological methods in cancer cure, the review concluded, was very disappointing, and there is little hope of development of a successful immunotherapy of cancer [23 p. 132, 192].

In the 1930s and 1940s, studies made with the inbred strains of mice developed at the Jackson Memorial Laboratories, Bar Harbor, Maine, confirmed the strong suspicion that phenomena previously gathered under the heading "resistance to transplanted tumors" were in fact manifestations of the body's resistance to a genetically-different tissue [25]. In 1942 Dr. Murphy used inbred strains of mice to prove the hypothesis that "acquired resistance to tumors" was similar to "natural resistance to tumors". He was able to show that so-called acquired resistance to tumors – an accelerated rejection of a grafted tumor by an animal injected with living cells from another individual belonging to the same species – could not be observed when both the donor and the recipient of the cells belonged to the same inbred strain. The "resistance to malignant growth" was, Dr. Murphy proposed, a laboratory artifact, born of confusion between a "resistance" to tumors and a rejection of genetically different tissue [26]. Lymphocytes might well play a role in the "resistance" phenomenon, but such "resistance" was of no interest for the oncologist looking for a cure for human malignancies.

3.3 Immunology and Cancer –
The Search for Tumor Antigens

During the "chemical period of immunology", immunologists focused on studies of specific antibodies and antigens [11]. Accordingly, in the 1940s and 1950s studies of links between cancer and immune mechanisms were focused on chemical investigations. Researchers looked for chemical structures which were specific to the malignant tissue and were not expressed by normal cells. Such "cancer antigens", if discovered, would allow early diagnosis of malignant tumors and would facilitate the development of immunological methods to prevent and fight cancer. The search for the putative "tumor antigens" attracted specialists in biochemistry and immunochemistry to cancer studies. These specialists introduced new techniques for studying proteins (ultracentrifugation, electrophoresis, radiolabeling and fluorescent labeling) which allowed for more detailed studies of antigenic structures (that is, chemical structures which are recognized as "foreign" by the body and are able to elicit the formation of specific antibodies). They were not very successful. A review of the state of the art in cancer immunology – written by Dr. Hauschka in 1952 – reported numerous studies which applied the latest biochemical and immunochemical methods to the search for putative cancer antigens. The importation of new biochemical and immunochemical methods into a medically-oriented field did not lead, however, to new developments. Ultramicroscopic investigations and chemical studies (studies of ribosome density, nucleic acids, levels of intracellular enzymes) did reveal quantitative differences between normal and malignant cells, but all these differences could be explained by the rapid rate of multiplication of malignant cells. In contrast, Dr. Hauschka explained, extensive studies did not allow the identification of tumor-specific antigens in naturally occurring tumors [27]. The failure to find such tumor-specific antigens cast doubt over the possibility of developing efficient immunotherapies of cancer. Dr. Chester Southam, who reviewed the field of experimental tumor immunology in 1960, similarly affirmed that the most basic problem "is whether host defense mechanisms against cancer actually exist." The hope of finding such mechanisms, Dr. Southam explained, had a reasonable basis. There are well-documented, although rare cases of spontaneous regression of cancer in humans and in laboratory animals, inflammatory cells can be found around some malignant tumors, and hosts with advanced cancer often have impaired immune functions. Researchers hoped that the perfection of immmunological methods of detecting even minute amounts of antigen in the 1950s would lead to a description of tumor-specific antigens, or to the demonstration of efficient involvement of immune mechanisms in the control of malignant growths. The introduction of these new methods did not lead, however, to new findings: Dr. Haushka's description of the state of the field in 1952 remained valid in 1960 too. The future of cancer immunology was, Dr. Southam concluded, uncertain [28].

3.4 Immunotherapy and Bedside Tinkering

Between 1895 and 1960 nearly every theoretically conceivable approach to application of immunological methods to cancer therapy was tried at one time or another on patients. Attempts at cancer immunotherapy were nearly always small-scale experiments, typically conducted by one or a handful of investigators interested in immunological methods, and usually motivated by a desire on the part of physicians to "do something" for otherwise incurable patients. These attempts were characterized by great heterogeneity in experimental design, in criteria of inclusion, and in patient follow-up. Even the most successful of these methods, namely the application of bacterial toxins to cancer therapy, never reached the stage of standardization and was applied by individual doctors in a variety of ways [29,30]. Bedside tinkering is not unusual in the clinics, and it may lead to the development of clinical innovations. This was not, however, the case with cancer immunotherapy. Nearly every method tested was presented by some experimenters as successful, but no one single approach became an accepted therapy for cancer.

The most successful early "immunotherapy" of cancer was based on injection of bacterial toxins. This therapy was born in the clinics, the fruit of an unexpected clinical observation. Dr. William Coley, a surgeon at the Memorial Hospital, New York, observed a spontaneous regression of residual, inoperable sarcoma in a patient suffering from a severe attack of streptococcal infection – erysipelas. Intrigued by this phenomenon, he screened the literature and found 38 earlier descriptions of regression of cancer following a bacterial infection. He decided to attempt to cure cancer by artificial induction of an attack of erysipelas. Such therapy was dangerous. Among the first patients artificially infected with erysipelas, one was cured and three died of complications following the infection. Dr. Coley thus shifted to using killed cultures of streptococci, later mixed with toxins of another microorganism, *Bacillus prodigiosus* (today *Serratia marcescens*), a mixture later named

"Coley's toxins". Dr. Coley tried his new method on a large number of cancer patients, and claimed to obtain positive results in numerous cases of inoperable sarcoma. He interpreted the antitumor effects of toxins in terms of stimulation of the body's "resistance" to cancer [31].

Other physicians confirmed that occasionally, and in an unpredictable manner, Coley's toxins induced regression of advanced cancers. The therapy remained controversial, but in the early twentieth century it was one of the few treatments that could be proposed to a patient suffering from inoperable cancer. Doctors were therefore willing to use this therapy as a last-ditch attempt to help their patients. Dr. Coley's therapy never reached the stage of systematic testing, was not regularly used in a major hospital (Dr. Coley's own department at the Memorial Hospital was the sole exception), and its application was never rigorously codified. Nevertheless, in the late nineteenth and early twentieth century individual doctors continued to employ Coley's toxins. Experiments with laboratory animals confirmed the antitumor effects of the toxins, but were unable to point to the possible mechanism of such effects [32]. In the meantime the efficacy of this therapy in patients remained low and its results were highly variable. (The difficulties of obtaining reproducible clinical results with Coley's toxins were attributed to the variability of commercial preparations of the toxins and to their inadequate use by physicians [33].) The advent, in the 1920s, of a new and prestigious therapy for inoperable cancer – radiotherapy – greatly diminished the clinical application of Coley's toxins.

Interest in Dr. Coley's therapy was temporary renewed in the late 1930s. The variability and lack of predictability of the anticancer activity of Coley's toxins, some researchers then reasoned, could perhaps be explained by the assumption that different preparations of the toxins contain variable amounts of unknown anticancer substance(s). It would therefore be interesting to attempt to isolate the active compounds of Coley's toxins and study their pharmacological properties. One such research program was initiated in the late 1930s by Dr. Shear and his colleagues at the U.S. Public Health Service Office of Cancer Investigations at Harvard University [34]. This project successfully linked laboratory and clinics, insofar as substances, such as the polysaccharide of *Serratia marcescens* – the active compound of "Coley's toxins" – were purified by chemists, tested in laboratory animals, and then in humans. It was not, however, an "immunotherapy" program but a "chemotherapy" one, because it aimed at finding molecules able to selectively kill tumor cells in the human body. When screening of substances isolated from bacteria, fungi and plants failed to uncover efficient anticancer drugs – for example, the polysaccharide of *Serratia marcescens* was found to be too toxic for clinical use – Dr. Shear's program shifted to large-scale tests of synthetic chemical compounds with potential anticancer activity. It became one of the first screening programs for chemotherapies of cancer, and the forerunner of the National Cancer Institute (NCI) chemotherapy programs [35]. In the meantime, sporadic attempts at immunotherapy of human malignancies continued in the 1940s and 1950s. However, these clinical experiments were not coordinated, the number of enrolled patients was usually small, patients often suffered from a vast array of tumors, and clinical evidence was mainly anecdotal. Dr. Southam, who reviewed in 1961 the area of cancer immunotherapy in the clinics, concluded that all these attempts had contributed nothing to applied immunology, had failed to give any indication of cancer-specific antigens in humans, and that there was no proof that they had ever influenced the course of cancer in a single patient [36].

3.5 The Use of Bacterial Vaccines as Cancer Therapy

The1960s were characterized by an exponential growth in both experimental cancer studies and in immunological research [37]. In the 1960s a growing number of cancer researchers turned to immunological studies, while numerous immunologists became interested in malignant tumors [38]. In the 1960s the practical success of kidney transplantation stimulated immunologists interested in the cellular mechanisms of graft rejection. The unraveling of the role of specific sub-sets of lymphocytes in the destruction of grafted "foreign" cells led to the hypothesis that similar cellular mechanisms may play a role in the killing of "deviant" cancer cells too. There was a sharp rise in the number of clinical trials of cancer immunotherapies in the late 1960s and early 1970s [39,40]. rise was probably related to a general increase of clinical trials of new anticancer therapies, and to the development of medical oncology as a trial-oriented professional segment. Several "old-new" methods of immunotherapies were tested again in the 1960s and 1970s, among them active immunotherapies (vaccination with tumor cells) and passive immunotherapies (the use of antitumor sera). The quantitative and qualitative transformation of the domain of cancer immunotherapy was, however, mainly the result of large-scale tests of non-specific active immunotherapy – that is, attempts at the non-specific stimulation of immune mechanisms, usually with bacterial vaccines or their derivatives. The stimulating agent most frequently used was a living, attenuated strain of the tuberculous bacillus, the BCG (*Bacillus Calmette-Guerin*), a microorganism found to stimulate cellular immune responses in laboratory animals [41].

In the mid-1960s a large group of French investigators, led by the hematologist Dr. Georges Mathé, started to treat leukemia patients with a combination of BCG and specific tumor vaccines. Dr. Mathé had a long-standing interest in experimental studies of hematological cancers, and was one of the pioneers of clinical trials of drug treatment of leukemia in France. In addition, Dr. Mathé had a part in the widely publicized – and partially successful – attempt to transplant bone marrow in order to save the life of an accidentally irradiated Yugoslav atomic scientist in 1958 [42]. The combination of familiarity with experimental cancer research, with organization of chemotherapy trials and with the practice of transplantation of tissues, provided a fertile ground for the development of a new approach to immunotherapy of cancer. Dr. Mathé's main aim in his immunotherapy trial was the prolongation of chemotherapy-induced remission in leukemia patients. The results of the first series of experiments published by Dr. Mathé and his collaborators were very impressive. Patients suffering from acute immunoblastic leukemia and treated with a combination of BCG and specific cancer vaccines, had much longer remissions and lived significantly longer than controls treated with chemotherapy alone [43–45]. The advocates of immunotherapy claimed that in the majority of cases chemotherapy alone was not sufficient to get rid of disseminated disease, and that only a parallel stimulation of immune mechanisms could lead to the elimination of residual malignant cells [46].

The publication of the results of Dr. Mathé's group, and the claims made by Gutterman and his collaborators as well as by Morton and his associates, that treatment with BCG prolonged the survival of patients with malignant melanoma, opened a period of enthusiasm for cancer immunotherapy [47,48]. A NCI-sponsored international registry of immunotherapy trials recorded 347 immunotherapy trials in 1976, mostly using non-specific bacterial stimulants: 198 trials used BCG as the stimulating agent, 38 a BCG fraction MER (methanol extract residue), 69 another microorganism, *Corynebacterium parvum*, and only 8 used specific tumor vaccines [49]. The main difference between the attempts at immunotherapy of cancer made in the 1970s and earlier efforts to stimulate immune mechanisms in cancer patients, was not technical but organizational – the switch to large-scale multicenter clinical trials and a successful cooperation with the laboratory. The rapid development of cellular immunology in the 1960s and 1970s, and growing integration of clinical immunology laboratories into hospitals, enlarged the possibilities of monitoring the patients' immune responses. Many investigators believed that the introduction of new immunological methods would radically transform the practice of cancer immunotherapy, moving it from a "pre-scientific" to a "scientific" era [50].

In the 1970s, experimental tumor immunology became an important domain of biological investigation. The accumulation of experimental data was not translated, however, into better clinical results. First optimistic reports on the impressive clinical results of immunotherapy in leukemia were not duplicated in larger multicenter clinical trials organized by the British Medical Council. Immunotherapy and chemotherapy were not found to be superior to chemotherapy alone [51]. Dr. Mathé's angry answer was that the conditions of the adequate duplication of his experiments had not been met [52]. Dr. Mathé's protests notwithstanding, in the late 1970s the majority of investigators agreed that the efficacy of BCG in leukemia therapy was not proven [40,53]. Similarly, researchers failed to reproduce studies which claimed that BCG therapy improved the clinical status of patients with solid tumors (one notable exception was the proven efficacy of BCG in the treatment of superficial cancer of the bladder [54–57]). Even partisans of the new therapy were obliged to conclude that "immunotherapy as presently practiced, seems to be a relatively ineffective way to treat human tumors." [58]. They continued, however, to express their faith in the future of this therapeutic approach [56]. The same professional groups that introduced immunotherapy to the cancer clinics in the 1970s – medical oncologists and immunologists – became, from the mid-1970s on, involved in the development of a different approach to immunotherapy of cancer – the use of cytokines (at first interferon, and then interleukin-2). They were joined by a new group – molecular biologists. In addition, immunotherapy of cancer was shaped in the 1980s by a new and important development: the alignment of the interests of immunologists, medical oncologists and molecular biologists, with those of the biotechnology industry.

3.6 Interleukin-2: From the Test Tube to Patients

In the 1970s, immunotherapy of cancer was viewed as the undisputed domain of the immunologist, because studies in this area used materials (bacterial vaccines, bacterial products) and methods (skin tests, measures of levels of antibodies in the blood, studies of activation of immunocompetent cells) firmly rooted in the immunological tradition [37,59]. However, the introduction in the 1970s of a "natural" regulatory molecule – the interferon – to cancer therapy, complicated the debate on "ownership" of the new approaches to therapy of cancer. Interferon was first credited with antiviral properties, and only later was one type of interferon molecule credited with the ability to enhance immune responses. It was thus defined as "biotherapy", rather than "immunotherapy" of cancer. In the late 1970s interferon focused the hopes for a highly efficient "biological therapy of cancers" that, unlike other treatments for this disease, would provide cures without toxicity. These hopes never materialized. Interferon had demonstrable antitumor effects in humans, but this molecule seldom produced dramatic therapeutic results and it induced important side effects [60–64]. While interferon was not found to be "the penicillin for cancer", it had non-contested therapeutic effects in selected cancers. It ended by being included in the therapy for a growing number of malignancies, and in the early 1990s it became a profitable product for the biotechnology industry [65].

Circa 1983, experts had turned to another cytokine which held the promise of antitumor activity, interleukin-2 (IL-2). IL-2 was known as a molecule that stimulates the growth of T-lymphocytes, a sub-population of lymphocytes which play an important role in numerous immune mechanisms (hence the earlier name, T-Cell Growth Factor or TCGF [68]). The TCGF, renamed interleukin-2 or IL-2 in the early 1980s, when this molecule was included in a group of substances which transmit messages between white blood cells (and therefore called "interleukins"). IL-2 made the culture of specific sub-populations of T-cells in the test tube possible. For that reason it rapidly became an important research tool for cellular immunologists. In parallel, the IL-2 molecule was immediately perceived as having important therapeutic potential, and it attracted the attention of the pharmaceutical industry, interested in regulatory molecules. In 1983 the gene for IL-2 was cloned independently by several groups of molecular biologists: T. Taniguchi's group in Tokyo, Japan (collaborating with the Japanese firm Ajinomoto), Robert Gallo's group in Bethesda, Maryland, USA (linked with the Genetics Institute), R. Fiers's group in Gand, Belgium (collaborating with Biogen), and R.J. Robb's group in Wilmington, Delaware, USA (linked with Dupont de Nemours [69]). Circa 1980, when the studies that led to the cloning of IL-2 had begun, this molecule was not perceived exclusively – or even primarily – as an anticancer treatment. Researchers looked for clinical applications of IL-2 in immune deficiency states (that is, in cases of inefficient functioning of immune mechanisms), in particular those in which there was a dysfunction of T-cells. Such dysfunction, the specialists claimed, was often observed in advanced cancer, and it was not illogical to consider the administration of IL-2 to cancer patients. This therapeutic application of IL-2 was, however, viewed as but one possibility among many others (the therapy of AIDS and of inborn immunodeficiencies were more obvious candidates for treatment with a molecule which stimulated T-cells). The shift towards the perception of IL-2 as a biological therapy for cancer was influenced by the activity of a particularly visible and dynamic group of oncologists – Dr. Steven Rosenberg's group at the Surgery Branch of NIH – and by the privileged relationship between that group and the Californian biotechnology firm Cetus.

Dr. Rosenberg, a surgeon who did a PhD in biophysics before completing his surgical residency, wished to combine the professional roles of pre-clinical researcher and medical practitioner. He was appointed head of the Surgery Division of the NIH because the NIH directorship was interested in the introduction of modern biological research methods into that traditional, skill-oriented division [70]. In the 1960s Dr. Rosenberg was involved in last-ditch attempts at adoptive cancer immunotherapy in terminally ill cancer patients [71]. In the 1970s Dr. Rosenberg took part in clinical trials of BCG in melanoma and bone sarcoma, and in 1982 he concluded that bacterial vaccines did not show clinical efficacy against cancer [53]. He remained convinced, however, that immune mechanisms might play an important role in the destruction of malignant cells. In the early 1980s Dr. Rosenberg became interested in "adoptive immunotherapy" – the transfer to the patient of white blood cells which had acquired an antitumor activity in the test tube. This therapy, he believed, should have multiple advantages. It should have low toxicity, because immunized cells should be capable of attacking tumor but not normal tissues, it should not limit the efficacy of natural immune mechanisms, and finally, this therapy should be easy to combine with the existing cancer treatments [72].

Adoptive immunotherapy was first introduced into treatment of human malignancies in the 1950s and 1960s, following the observation that cytotoxic lymphocytes are able to destroy malignant cells in the test tube. In early attempts at adoptive immunotherapy, patients received leukocytes from volunteers vaccinated with the patient's own tumor cells. The failure of these attempts was usually attributed to the difficulty of generating adequate amounts of efficient cytotoxic cells. The discovery of a family of molecules (cytokines, growth factors) which would make possible the large-scale culture of specific sub-sets of white cells in the test tube, opened new horizons for the adoptive immunotherapy of cancer. IL-2 was found to be a particularly effective tool for expanding selected populations of lymphocytes [73]. Dr. Rosenberg and his collaborators aimed first at the development of tumor-specific adoptive immunotherapy with IL-2-activated tumor-specific cytotoxic T-cells. When they incubated blood cells with IL-2, they noticed that under these conditions normal human leukocytes spontaneously acquired the ability to lyse tumor cells [74]. The lysis of tumor cells was non-specific, and was not related to an immunization process. Dr. Rosenberg and his co-workers were, however, interested above all in finding an efficient way to destroy malignant tumors, and were not attached to their initial hypothesis of a specific "adoptive immunotherapy". Their observation became the starting point of several years of efforts to introduce IL-2-activated lymphocytes into cancer therapy.

Dr. Rosenberg's group arrived at the conclusion that IL-2 activated a previously unknown sub-set of cells, which they named leukin-activated killer cells or LAK cells. LAK cells, Dr. Rosenberg and his co-workers proposed, were a previously unknown subset of highly efficient cytotoxic lymphocytes which would be especially good candidates for adoptive cancer immunotherapy in man [75,76]. In order to obtain permission to experiment on humans, it was first necessary to demonstrate the therapeutic effects of LAK in tumor-bearing laboratory animals. Dr. Rosenberg and his colleagues rapidly switched to studies of the therapeutic properties of LAK cells in a mouse model which, Dr. Rosenberg explained later, faithfully mirrored the clinical situation of advanced, disseminated cancer [72]. They were indeed able to demonstrate the regression of disseminated tumors, but this effect was obtained only in mice injected with very elevated concentrations of LAK cells, and was more pronounced if mice received intravenous injections of IL-2 together with the LAK cells [727778].

In articles which reviewed the development of LAK therapy, Dr. Rosenberg first described the lysis of malignant cells by LAK cells in the test tube, then the elimination of experimentally-induced tumor metastases in mice by LAK transfusion and by the combination LAK and IL-2, and finally the first experiments in humans [79]. The narrative sequence creates the impression that IL-2 studies followed an orderly path from the test tube, through experiments in laboratory animals, to experiments on humans. The chronology of publications of Dr. Rosenberg's laboratory shows, however, that Dr. Rosenberg and his collaborators simultaneously pursued studies in the test tube, experiments with animals, and preliminary clinical investigations in humans. All these studies seem to be driven by the practical goal of developing rapidly an efficient adoptive immunotherapy of human cancer, and by a strong belief in the feasibility of such therapy [80]. They were also stimulated by the knowledge that enough recombinant IL-2 would be available for human experimentation, thanks to the collaboration (which started in 1982) between Dr. Rosenberg's laboratory and the biotechnology firm Cetus [81]. The combination of strong faith in the antitumor efficacy of IL-2 activated cells access to a nearly unlimited supply of IL-2, and pressure to obtain exploitable clinical results, may have guided Dr. Rosenberg's rapid switch to the use of high, non-physiological (thus potentially toxic) doses of IL-2 in his attempts to cure cancer in mice, and then in man.

The first experiments in humans started in 1983 (7 cancer and 5 AIDS patients participated in this trial [82]). The purpose of these experiments was to obtain preliminary information on the physiological effects of IL-2, and to compare the biological effects of cell-secreted IL-2 to those of recombinant IL-2 (IL-2 produced by genetic engineering techniques; [83]; 20 cancer patients participated in this trial). In these experiments the injection of IL-2 into patients suffering from advanced cancer did not modify their clinical status. A second series of experiments attempted to establish the maximum tolerated dose of recombinant IL-2. It was found that high concentrations of IL-2 provoked serious side effects: fever, chills, malaise, intestinal symptoms, muscle and bone pain and finally a "capillary leak syndrome" – the escape of water from small blood vessels which induced weight gain, edema, and sometimes severe respiratory and circulatory complications. (Toxic effects were also observed in laboratory animals [84].) This last effect of IL-2 was unexpected, because the "capillary leak syndrome" was not observed in mice. However, the observation that IL-2 had a high degree of toxicity in humans did not delay the beginning of the second stage of the study – therapy with LAK cells and high concentrations of IL-2. That stage was guided by the assumption that the clinical efficiency of the new therapy would be directly proportional to the dose of IL-2/LAK used. Clinical trials of combined LAK/IL-2 therapy using the highest tolerated dose of IL-2 together with LAK cells had started in the early spring of 1985. Dr. Rosenberg himself later described the way he "pushed" his patients to the upper limits of their physical tolerance, and sometimes even to the brink of death with cardiac and respiratory problems. This "heroic" approach was successful. In spring 1985 Dr. Rosenberg and his colleagues witnessed their first cure: a spectacular, partial regression of pulmonary metastases in a melanoma patient, then the complete (and stable) regression of subcutaneus nodules in another melanoma patient [70].

In the summer of 1985 there was a widespread rumor in the oncological milieu that Dr. Rosenberg had developed a new therapy for previously incurable advanced solid tumors. In October 1985 Dr. Rosenberg received the General Motors Cancer Research Award. The new cancer treatment was discussed in the November 1985 issue of *Fortune Magazine*, and the news had a favorable effect on the shares of biotechnology firms involved in cloning and producing IL-2 for clinical use [85]. In December 1985 an article describing the first results of the new therapy was published in the *New England Journal of Medicine* [86], and the discovery of a new cancer biotherapy was officially announced by the National Cancer Institute. Magazines had run cover stories about the new cancer cure, and the announcement of the first results of the LAK cell therapy experiments was also carried by the major American television networks. *The Wall Street Journal* strongly recommended de-regulating the FDA rules of approval for new drugs in order to speed up patients' access to promising therapies such as LAK/IL-2 [87], while NCI director Dr. Vincent de Vita Jr. explained that the new treatment "is the most interesting and exciting biological therapy we've seen so far" [88]. (On the enthusiastic reception of the announcement of the new cancer theapy by the media see Patterson [89]). This statement was based on the first clinical results reported by Dr. Rosenberg: the tumors of 11 of the 25 patients suffering from advanced, incurable cancer shrank by 50% or more following IL-2/LAK therapy. The new treatment induced severe toxicity and in many cases it required close monitoring of the patients in an intensive care unit, but according to Rosenberg, all the toxic effects of the new therapy disappeared when IL-2 administration was discontinued. IL-2 treatment looked very promising indeed.

3.7 IL-2 as an Anticancer Drug

The excitement induced by Dr. Rosenberg's 1985 article led to a rapid diffusion of clinical trials of IL-2/LAK in the United States and abroad. An American philanthropist, Dr. Armand Hammer, donated $150,000 to the NCI for further research on this subject, and the NCI allocated 2.5 million dollars for six additional extra-mural (non-NCI) studies of IL-2/LAK treatment [90]. In addition, a clinical trial of IL-2/LAK therapy was initiated in 1985 by physicians linked with Biotherapeutic Inc., a private institute dedicated to patient-funded experimental cancer therapies [91]. The most important innovation introduced by Biotherapeutic specialists was that IL-2-treated patients received continuous intravenous infusion of interleukin, not, as in the NCI clinical trial, bolus injections of this molecule. The new method of administration of IL-2, Biotherapeutic Inc. researchers claimed, diminished the side effects of IL-2 therapy and enhanced its safety without limiting its effectiveness. Out of 40 IL-2 treated patients 15 responded to this treatment [92].

The encouraging news that a different way to administer IL-2 might limit the toxic effects of the new treatment were moderated by the finding that early evaluations of IL-2/LAK treatment were too optimistic. In 1986 a study by Dr. Rosenberg's group indicated that IL-2/LAK therapy, but also high doses of IL-2 alone, have antitumor effects. But it also indicated that the rate of therapeutic success – obtained in a much larger sample of patients – was significantly lower than the one reported in the first article published by Dr. Rosenberg's group [93,94]. Intermediary reports of the six extra-mural clinical trials of IL-2/LAK therapy also pointed to the relatively low efficacy of the IL-2/LAK treatment [95]. These findings were particularly disappointing because the new clinical trials were centered exclusively on "responding" tumors: melanoma and renal cell carcinoma. Lower than expected efficacy is not, however, synonymous with the absence of efficacy, and IL-2/LAK treatment did show undeniable effects on otherwise untreatable cancers. On the other hand, the interleukin treatment was expensive, it induced severe, sometimes life-threatening toxic effects, and only in a small proportion of treated patients was its physiological effect – the reduction of the volume of a tumor – translated into a stable clinical result – prolonged remission of malignant disease. The publication of the results of numerous additional clinical trials of IL-2/LAK and IL-2 therapy in the years 1987-1990 confirmed that in IL-2 therapy 15–25% of the patients suffering from "sensitive" tumors, melanoma and renal cell carcinoma, reacted to the treatment, that is, that the total volume of their tumors was reduced by 40% or more. Among these respondents there was a small number of long-term remissions of otherwise untreatable malignancies (18 among the 652 patients included in this trial had such long-term remissions, and in 14 patients the remission lasted for more than a year). This trial also confirmed that therapeutic doses of IL-2 generated severe side effects [96–98]. In particular, a large proportion (between one-third and one-half) of the patients suffered from serious neuropsychiatric perturbations, attributed to physiological effects of IL-2 therapy such as capillary leak syndrome, impaired blood circulation and non-specific liberation of chemical substances by IL-2- stimulated cells. The symptoms disappeared when the therapy was discontinued [99,100].

Circa 1989, cumulative results of the clinical trials of IL-2/LAK pointed to the absence of important differences between clinical results obtained with and without LAK cells [101–103]. Oncologists were often relieved to find that LAK cells were not indispensable for IL-2 therapy. The necessity of producing these cells increased the costs of this therapy and made it more labor-intensive. At the same time, the argument that IL-2's antitumor activity was mediated through the induction of high levels of LAK cells in the body, was not confirmed by experimental findings. No significant correlation was found between the level of "LAK activity" in the blood of IL-2 treated patients and their clinical responses to interleukin therapy [104,105]. In the absence of a clear-cut correlation between the levels of activated LAK in the blood of IL-2 treated patients and clinical results of IL-2 therapy, some experts proposed that a migration of activated LAK to the tumor site might explain why the increase in the activity of LAK cells was undetectable in circulating blood [104–106] (The increased expression of MHC antigens, some investigators proposed, may facilitate the recognition and the destruction of malignant cells by immune mechanisms). Other investigators proposed that different immunological mechanisms, such as the stimulation of secretion of other cytokines, or of expression of major histocompatibility complex antigens on tumor cells, might account for the antitumor effects of IL-2. The problem was summed up in 1989 by a pioneer of IL-2 studies and one of the leading specialists in the field, Dr. Kendall Smith: "We still do not know which biochemical pathways the interleukins activate." [107].

In the late 1980s investigators began to exploit new avenues for the uses of IL-2 as an anticancer drug. Here again, Dr. Rosenberg's laboratory led the way. Dr. Rosenberg's new efforts were at first guided by his long-standing aspiration to develop an adoptive cancer immunotherapy. It was possible, he speculated, that LAK cells were not very efficient in the organism because they were not sufficiently tumor-specific. Dr. Rosenberg therefore returned to his – never entirely abandoned – original idea to use IL-2 merely as a technical device which allowed the expansion of specific sub-populations of cytotoxic lymphocytes in the test tube. Tumors, Dr. De Fano had observed as early as 1912, are often infiltrated by lymphocytes [108]. Dr. Rosenberg and his collaborators assumed that it is probable that the lymphocytes which spontaneously enter a malignant growth – which they named tumor infiltrating lymphocytes (TIL) – contain a significant proportion of specific cytotoxic cells directed against the tumor. TIL were a heterogeneous population composed mainly (but not exclusively) of cytotoxic T-lymphocytes. Dr. Rosenberg and his collaborators started to study TIL in 1980 [109]. The observation that IL-2 activated peripheral blood lymphocytes (a cell population which is much easier to obtain than tumor infiltrating cells) may develop non-specific antitumor activity when cultured in the presence of IL-2, shifted their attention to LAK cells. Later, however, the relatively low clinical efficacy of LAK cells brought them back to the idea of expanding selected populations of autologous tumor infiltrating lymphocytes in the test tube, then injecting these cells into patients.

In 1986 Dr. Rosenberg's group developed a method to isolate the lymphocytes which infiltrate human melanoma [110]. At the same time, they found that TIL were much more efficient than LAK cells in animal models [111] (The murine tumor models employed in this study were similar to those used to demonstrate the antitumor activity of LAK). Dr. Rosenberg's group also affirmed that antitumor effects of TIL could be observed in mice even in the absence of IL-2, and were optimized by the addition of much lower doses of IL-2, as compared to the optimal activation of LAK cells. The main drawback of the new method was the difficulty of preparing TIL. These cells were cultivated from surgically-excised tumors incubated in the presence of IL-2. The tumor cells gradually died in culture, while the tumor infiltrating lymphocytes multiplied in the presence of interleukin-2. The number of TIL obtained using this method was, however, small. A costly and labor-consuming long-term culture (in human tumors, 6-8 weeks, sometimes even more) was necessary to obtain enough cells for an adoptive immunotherapy. Nevertheless, the method seemed promising enough to start experiments in humans.

The first series of patients treated with TILs at the NCI was composed of melanoma patients only, because tumor infiltrating lymphocytes were particularly abundant in melanoma, and it was easy to excise subcutaneous nodules of that tumor. The patients received autologous TIL, and – surprisingly if one takes into account the published results of animal studies – a full dose of IL-2 (that is, the maximally tolerated dose of that molecule). The IL-2 injection induced the same undesirable effects as during a standard IL-2/LAK therapy. These effects were, however, usually less severe because the TIL treatment was shorter. Among the first 20 patients, 9 out of the 15 who were not treated previously by IL-2 and LAK responded to the TIL therapy, while 2 of 5 patients who did not respond to previous therapy with IL-2 responded to the new therapeutic regimen. These impressive results were moderated by the observation that only one patient had a complete remission and remained disease-free after 13 months, while other remissions were partial and lasted 2-9 months [112].

Researchers in the Rosenberg group devised two approaches to improve the results of TIL therapy: the combination of TIL treatment with other biological response modifiers, and the genetic manipulation of TIL. The first approach entered early clinical stages circa 1991. TIL were injected together with other substance(s) that modulate the growth and function of lymphocytes, such as interleukin-4, interleukin-6, interferon and antibodies against transforming growth factor-beta (TGF-ß). Another variant of that approach was an attempt to generate more efficient TIL in the test tube through selective culture of tumor-specific clones of cytotoxic T-cells in the presence of different growth factors [113]. The second solution tested by Dr. Rosenberg's group was the genetic modification of TIL and the development of "super-TIL" by inserting into "normal" TIL genes that would greatly increase the tumor-killing efficacy of these cells (e.g., genes that code for cytotoxic molecules, such as the tumor necrotic factor [71]). In the first phase of the new project, Rosenberg and his collaborators injected patients with TIL containing a neutral marker (bacterial genes resistant to neomycin). The modified cells were found in the blood up to 2 months after their administration, and were also recovered from tumor deposits proving that at least some of the TIL migrate to the tumor site [114]. Dr. Rosenberg at first presented the new project as a continuation of his efforts to develop more efficient cytotoxic cells, but he also stressed the potential importance of genetically-modified lymphocytes in the therapy for genetic diseases [115].

While Dr. Rosenberg and his collaborators had taken IL-2 treatment as a starting point for innovative therapeutic approaches, other investigators attempted to improve the therapy through more traditional means. Physicians attempted to simplify the new method by reducing the dose of IL-2 administered to patients [116,117] (in some studies, low doses of interleukin were injected subcutaneously into cancer patients, and the authors of these studies affirmed that IL-2 administered in that way was clinically and immunologically active [118,119]). Claims that it is possible to use lower (and therefore less toxic) doses of IL-2 results were, however, viewed with scepticism by the majority of the promoters of interleukin therapy, who maintained that the best clinical results were obtained with the highest tolerated doses of IL-2

[79,120]. Another approach, favored by many medical oncologists, was the combination of interleukin therapy and chemotherapy. These attempts were made mainly in melanoma (there is no known efficient chemotherapy for renal cell carcinoma [121–123]). Other trials experimented with the sequential combination of several treatments – chemotherapy, interleukin-2 and other cytokines – and attempted to find the best timing for the administration of this therapy [124,].

The methods employed in "post LAK" clinical trials of IL-2 therapy closely recall clinical trials of cancer chemotherapy: modulation of doses and administration methods of the active substance, attempts to find the "winning combination" of drugs, a search for the best timing for the treatment. This is not surprising. Interleukin-2 was "naturalized" by clinical oncologists, who integrated the new molecule into their routine therapeutic approaches and their "native culture" of experimentation in the clinics. A parallel effort at "naturalizing" IL-2 was undertaken by pharmaceutical firms. In the mid- and late 1980s numerous small biotechnology firms disappeared, while the bulk of production of recombinant "biological response modifiers" shifted to larger, well-established pharmaceutical firms [85]. These firms applied traditional pharmacological research strategies to substances produced through genetic engineering. The "naturalization" of IL-2 by pharmaceutical companies led to attempts at the chemical manipulation of the IL-2 molecule in order to obtain less toxic and/or more effective substances, and to efforts to develop more efficient ways to deliver that molecule to target cells [126–128]. In the meantime, IL-2 obtained a marketing permit for a specific indication – the treatment of advanced kidney cancer – and it is used occasionally to treat other cancers as well. IL-2 was granted a marketing permit in the US in 1992 for the therapy of renal cell carcinoma and in 1997 for the treatment of melanoma. It was granted a marketing permit in France for therapy of renal cell carcinoma only in 1989 [129]. In the early 1990s clinicians and pharmaceutical firms continued their search for more efficient ways to use interleukins in cancer therapy. These attempts were seen, however, as useful improvement of existing therapies rather than as the promise of a "therapeutic revolution" in oncology. A review of cancer immunotherapy published in 1992 is entitled "Cancer immunotherapy: Are the results discouraging? Can they be improved?". It concludes that although some useful modifications are indeed possible, "dramatic improvements of generally applicable immunotherapy cannot be expected without conceptual and technical innovations" [130]. A review, "Tumor immunology: false hopes – new horizons?", published in 1998, proposes a similar view. It ends with the statement that while understanding of the molecular events underlying tumor development, and of the cellular and molecular basis of immune responses continue apace, the ability to exploit this new knowledge in the prevention and treatment of malignant tumors has met with limited success [131]. It is interesting to note that among the immunotherapies listed in this review one can find IL-10 and IL-12 (and also the BCG therapy for superficial cancer of the bladder, presented as the most efficient immunotherapy for any human malignancy), but not IL-2. The latter is now usually described as a pleiotropic regulatory protein with multiple physiological effects that need to be uncovered through intensive pre-clinical and clinical testing [129].

3.8 Concluding Remarks: Immunotherapy of Cancer – from Bedside Tinkering to "Big Medicine"

The first attempts at immunotherapy of cancer were isolated trials made by individual doctors during a period in which oncology did not exist as an organized and codified area of medical activity. In the early twentieth century, clinical trials of immunotherapy were made by individual pathologists and were not submitted to organized control by the medical community. Typically, the early trials were isolated studies conducted by a small number of investigators in a single institution. By contrast the multicenter interleukin trials were made in the framework of highly organized, powerful and interdependent international communities of oncologists and experimental cancer researchers. The recent clinical trials of cancer immunotherapy were not merely bigger or more sophisticated than the earlier ones – they were also qualitatively different. These changes in clinical trials of immunotherapy were the result of a gradual process with a radical outcome. Clinical trials of immunotherapy of cancer, once a small "cottage industry", were transformed into a powerful modern industry. This qualitative change was the combined consequence of expansion of basic research in oncology and immunology, of the growing importance of clinical experimentation in oncology, and of the development of close connections between clinics and the biotechnology industry.

The large-scale integration of academic scientists within the so-called biomedical-industrial complex led to important changes in immunological and oncological research. The term "biomedical-industrial complex" describes a dense, closely-knit network of cognitive, psychological, social and political interests in which members of distinct social groups coexist in a "symbiotic" relationship based on mutual legitimization [132–135]. The existence of the "biomedical-industrial complex" is often directly reflected in the power structure of major medical schools, biomedical research centers, government funding agencies and big disease-centered charities. It is also mirrored in the current forms of experimentation in the clinics. It favors the funding of therapeutic innovations which efficiently articulate multiple interests: those of clinicians, basic scientists, industrialists, health administrators and politicians.

The development of the "biomedical-industrial complex" is a relatively new event. Close relationships between biologists, clinicians and industrialists existed much earlier, and may be traced to the birth of "scientific medicine" in the second half of the nineteenth century [136]. However, before the Second World War links between the pharmaceutical industry and biomedical research were seldom central to mainstream academic research and clinical practices (important exceptions existed, however, in particular in the drug industry, where a few leading firms in the 1930s grasped the importance of links with academic science [137,138]). Relationships between fundamental scientists, clinicians and industrialists underwent important transformations in the aftermath of the Second World War, and again in the 1980s. In the late 1940s and 1950s the conviction that the development of biological and medical research was a necessary precondition for the solution of major health problems led to a massive influx of public funds to academic biomedical research, but also strengthened selected industrial collaborative efforts (e.g., in the production of antibiotics, hormones, and anticancer drugs). The organizational developments brought about by massive testing of anticancer drugs from the 1950s on were the consequence of these changes. In the late 1970s and 1980s the new biotechnology industry made a large-scale transformation of central scientific values of biomedical investigators into marketable assets possible. The development of interferon and interleukin-2 and then the patterns of their testing in the clinics and their introduction into routine therapy of malignant tumors, were made in the new context of "big biomedicine" closely related to the industry, but also closely supervised by regulatory governmental agencies.

The transformation of interleukin-2 into a drug is also related to the loss of its status as a "different" cure for cancer. Interleukin-2 was presented to the public under the double image of a substance representing the latest achievement of"high-tech medicine", which makes possible the production of drugs through genetic engineering technology, but also as a "natural substance" able to boost the body's defense mechanisms against cancer. This double image – technological achievement and natural substance, probably accounts for at least part of this molecule's appeal to the public imagination in the mid-1980s. Cytokines have lost their mythical aura in the late 1990s. Once presented in the media as a possible "penicillin for cancer" – they were integrated into the large spectrum of treatments used in the cancer clinics, and they are, as a rule, combined with other therapeutic approaches. This change in the status of cytokines is reflected in the subtitle of this symposium: "the dawn of recombinant protein drugs". This subtitle indicates only that the new substances are "drugs" – that is, physiologically active molecules – and that these drugs are "recombinant proteins " – that is, are manufactured using the methods of genetic engineering. It does not explain – as earlier publications did – that these drugs are a "fourth mode of cancer treatment", qualitatively different from the previous ones: surgery, radiotherapy and chemotherapy . In the 1980s, the early reports on the antitumor properties of interferon led to the development of NCI's program for the study of "biological response modifiers", seen as a new family of compounds which could be used in the treatment of human malignancies [139]. Today cancer experts tend to view all pharmacoactive substances as "biological response modifiers". Recently, the growing interest in therapeutic possibilities of cytokine antagonists – small molecules able to block cytokine receptors which can be elaborated through the practice of computer-assisted "rational drug design", displays the growing blurring of boundaries between the "biotherapy" and the "chemotherapy" of cancer. The loss of the labels "natural substance" and "immunotherapeutic agent" thus became part of the process of the naturalization of IL-2 in the clinics, and its (today still largely potential) transformation into a polyvalent therapeutic substance [129]).

References

1. Löwy I (1996) Between bench and bedside. Harvard University Press, Cambridge
2. Cohn S, Pick E, Oppenheim JJ (eds) (1979) The biology of lymphokines. Academic, New York
3. Herberman RB (1980) Natural cell-mediated immunity against tumors. Academic, New York
4. Rather JL (1978) The genesis of cancer. Johns Hopkins University Press, Baltimore
5. Rous P (1911) Transmission of a malignant new growth by means of cell-free filtrate. JAMA 56:198
6. Gaudillière JP (1993) NCI and the spreading genes: about the production of viruses, mice and cancer. Sociology of Science yearbook
7. Metchnikoff E (1901) L'Immmunité dans les maladies infectieuses. Mason, Paris, pp 283–284
8. Wright A, Douglas S (1903) An experimental investigation of the role of body fluids in connection with phagocytosis. Proc R Soc Lond 72:357
9. Duclaux E (1898) Traité de microbiologie, vol 1. Masson, Paris, pp 727–730
10. Silverstein AM (1991) The dynamics of conceptual change in twentieth century immunology. Cell Immunol 132:515–531
11. Silverstein AM (1988) Introduction. In: Bibel D (ed) Milestones in immunology. A historical exploration. Springer, Berlin Heidelberg New York, pp xiii-xv
12. Murphy JB, Rous P (1912) The behaviour of chicken sarcoma implanted in the developing embryo. J Exp Med 15:119–132
13. Murphy JB (1912) Transplantability of malignant tumors to the embryos of foreign species. JAMA 59:874–875
14. Murphy JB (1913) Transplantability of tissues to embryo of foreign species. J Exp Med 17:482–492
15. Murphy JB, Morton JJ (1915) The effects of X-rays on the resistance to cancer in mice. Science 42:842
16. Murphy JB, Morton JJ (1915) The effects of X-rays on the rate of growth of spontaneous tumors in mice. J Exp Med 22:800–803
17. Newspaper clippings from August 1915, Murphy's Papers, BM 956, American Philosophical Society, Philadelphia
18. Patterson J (1987) The dread disease: cancer and the modern American culture. Harvard University Press, Cambridge, pp 56–86
19. Murphy JB (1926) The lymphocyte in resistance to tissue grafting, malignant disease and tuberculous infection: an experimental study. The Rockefeller Institute for Medical Research, New York

20. Murphy JB (1915–1922) Reports of the directors of the laboratories, vol 4, 1915–1916, pp 226–227; vol 5, 1917, p 108; vol 6, 1918, p 54; vol 9, 1921, p 48; vol 10, 1922, p 101
21. Flexner S (1920) Report to the corporation, October 1920. In: Reports of the directors of the laboratories, vol 8, p 208
22. Austoker J (1988) History of the Imperial Cancer Research Fund, 1902–1986. Oxford University Press, Oxford
23. Wooglom W (1929) Immunity to transplantable tumors. Cancer Rev 4:124–195
24. Silverstein A (1989) A history of immunology. Academic, San Diego, pp 275–304
25. Barret MK (1940) The influence of genetic constitution upon the induction of resistance to transplantable tumors. J Natl Cancer Inst 1:387–393
26. Murphy JB (1942) An analysis of trends in cancer research. JAMA 120:107–111
27. Hauschka TS (1952) Immunologic aspects of cancer: a review. Cancer Res 12:615–633
28. Southam CM (1960) Relations of immunology to cancer: a review. Cancer Res 20:271–291
29. Coley Nauts H (1991) Bibliography of reports concerning the clinical or experimental use of Coley toxins (*Streptococcus pyogenes* and *Serratia marcescens*) 1893–1991. Cancer Research Institute, New York
30. Coley Nauts H (1989) Coley's toxins – the first century. Communication read at the meeting of International Clinical Hyperthermia Society, Rome, May 1989
31. Coley WB (1898) The treatment of inoperable sarcoma with the mixed toxins of erysipelas and *Bacillus prodigiosus*. Immediate and final results in 160 cases. JAMA 31:389–395, 456–465
32. Beebe SP, Tracey M (1907) The treatment of experimental tumors with bacterial toxins. JAMA 49:1493–1498
33. Coley Nauts H, Swift WE, Coley BL (1946) The treatment of malignant tumors by bacterial toxins and developed by the late William B. Coley M.D., reviewed in the light of modern research. Cancer Res 6:205–216
34. Zubord CG (1984) Origins and development of chemotherapy research at the National Cancer Institute. Cancer Treatment Rep 68(1):9–19
35. Bud RF (1978) Strategy in American cancer research after World War II: a case study. Soc Stud Sci 8:425–459
36. Southam C (1961) Applications of immunology to clinical cancer. Past attempts and future possibilities. Cancer Res 21:1302–1316
37. Good RA (1976) Runestones in immunology: inscription of journeys in discovery and analysis. J Immunol 117:1417–1429

38. Milder JW (1961) Introduction. In Proceedings of the symposium, The possible role of immunology in cancer, Rye, New York, March 16–18, 1961. Cancer Res 21:1169
39. Mastrangallo MM, Berd D, Bellet RE (1976) Review of immunotherapeutic studies in cancer patients. In: Martin M, Dionne L (eds) Immunocancerology in solid tumors. Stratton Intercontinental Medical Books, New York, pp 155–177
40. Terry WD (1977) Present status and future directions for cancer immunotherapy. In: Yamama Y, Kitagawa M, Azuma I (eds) Cancer immunotherapy and its immunological basis. Japan Scientific Societies Press, Tokyo, pp 239–246
41. Old LJ, Clarke DA, Benacerraf B (1959) Effect of Bacillus Calmette-Guérin infection on transplanted tumors in the mouse. Nature184:291
42. Mathé G, Amiel JL, Cattan A (1958) Transfusions et greffes de moelle osseuse homologue chez des humains irradiés à haute dose accidentellement. Rev Franc Etudes Clin Biol 4:3
43. Mathé G, Amiel JA, Schwartzenberg L et al (1969) Active immunotherapy for acute immunoblastic leukemia. Lancet i:697
44. Mathé G, Pouillart P, Schwartzenberg L et al (1974) Attempts at immunotherapy of 100 acute leukemia patients: some factors influencing results. In: Mathé G, Weiner R (eds) Investigation and stimulation of immunity. Springer, Berlin Heidelberg New York, pp 434–448
45. Mathé G, Amiel JA, Schwartzenberg L et al (1975) Immunothérapie active de la leucémie aigue et du lymphosarcome léucemique: une étude de 200 cas conduite pendant 10 ans. Nouv Presse Med 4:1337
46. Mathé G (1974) Introduction. In: Mathé G, Weiner R (eds) Investigation and stimulation of immunity. Springer, Berlin Heidelberg New York
47. Gutterman JU, Mavligit G, Macbride C et al (1974) Active immunotherapy with BCG for recurrent malignant melanoma. Lancet ii:128
48. Morton DL, Eilber FR, Holmes EC et al (1974) BCG therapy of malignant melanoma. Summary of seven years of experience. Ann Surg 10:635
49. Windhorst D (1978) International registry of tumor immunotherapy: the contribution of an information service for a new specialty. In: Terry WD, Windhorst D (eds) Immunotherapy of cancer. Present status of trials in man. Raven, New York
50. Clark, RL, Hickey RC, Hersh EM (eds) (1978) Immunotherapy of human cancer. Raven, New York, p ix
51. Concord trial (1971) A preliminary report. BMJ 4:189–194
52. Mathé G (1976) Round table. In: Lamoreux G, Turcotte R, Portelance V (eds) (1976) BCG in cancer immunotherapy. Grune and Stratton, New York, p 383

53. Terry WD, Rosenberg S (eds) (1982) Immunotherapy of human cancer. Elsevier, North Holland New York
54. Lamoreux G, Turcotte R, Portelance V (eds) (1976) BCG in cancer immunotherapy. Grune and Stratton, New York
55. Martin M, Dionne L (eds) (1976) Immunocancerology in solid tumors. Stratton Intercontinental Medical Books, New York
56. Terry WD, Windhorst D (eds) (1978) Immunotherapy of cancer. Present status of trials in man. Raven, New York
57. LoBuglio AL (ed) (1980) Clinical immunotherapy. Dekker, New York
58. Mastrangelo MJ, Berd D, Bellet RE (1978) Limitations, obstacles and controversies in the optimal development of immunotherapy. In: Clark, RL, Hickey RC, Hersh EM (eds) Immunotherapy of human cancer. Raven, New York, pp 375–394
59. New initiatives in immunology: NIAID Study Group Report (1981) NIH publication no 81–2215, National Institute of Health, Bethesda, Maryland
60. Mazumdar PM (ed) (1989) Working out the theory. In: Immunology 1930–1980:essays on the history of immunology.Wall and Thompson, Toronto, pp 7–8
61. Sikora K (ed) (1983) The cancer problem. In: Interferon and cancer. Plenum, London, pp 7–8
62. Powledge TM (1984) Interferon on trial. Biotechnology 2: 214–228
63. Talmage JE, Fidler IJ, Oldham RK (1985) Screening for the immunological response modifiers. method and rationale. Nijhoff, Boston, pp 180–181
64. Fent K, Zbinden G (1987) The toxicity of interferon and interleukin. Trends Pharmacol Sci 8:100–108
65. Therre H (1990) Le marché américain de monokines, cytokines et autres facteurs de croissance. Biofutur, January, pp 46–47
66. Editorial (1990) Biotechnol News 9:4–8
67. Dodet B (1994) Cytokines in the clinics. Choose your weapon. Eur Cytokine Network 5:369–376
68. Morgan, DA, Ruscetti FW, R. Gallo R (1976) Selective in vitro growth of lymphocytes from normal human bone marrow. Science 193:1007–1008
69. Fradzelli D, Roskam W (1983) Interleukine-2 humaine. Biofutur 2:340
70. Rosenberg SA, Barry JM (1992) The transformed cell. Putnam, New York
71. Rosenberg SA (1990) Adoptive immunotherapy for cancer. Sci Am May:34–44
72. Rosenberg SA (1987) Adoptive immunotherapy of cancer using lymphokine-activated killer cells and recombinant interleukine-2. In: DeVita VT, Hellman S, Rosenberg SA (eds) Important advances in oncology, 1986. Academic, New York, pp 55–91
73. Gillis S, Smith KA (1977) Long term culture of tumor-specific cytotoxic T-cells. Nature 368:154–155

74. Lotze MT, Grimm EA, Mazumder A, Strausser A, Rosenberg SA (1981) In vitro growth of cytotoxic human lymphocytes: IV. Lysis of fresh and cultured autologous tumor by human lymphocytes cultured in T-cell growth factor (TCGF). Cancer Res 41:4420–4425
75. Rosenberg SA (1984) The adoptive immunotherapy of cancer: accomplishments and prospects. Cancer Ther Rep 68:244–255
76. Yang JC, Mule JJ, Rosenberg SA (1985) Characterisation of the murine lymphokine-activated killer precursor and effector cell. Surg Forum 36:408–410
77. Ettinghausen SE, Lipford EH, Mule JJ, Rosenberg SA (1985) Systemic administration of recombinant interleukin-2 stimulates in vivo lymphoid cell stimulation in tissues. J Immunol 135:1488–1497
78. Donohue JH, Rosenstein M, Chang AE et al (1984) The systemic administration of purified interelukin-2 enhances the ability of sensitized murine lymphocyte lines to cure a disseminated syngeneic lymphoma. J Immunol 132:2123–2128
79. Rosenberg SA (1988) The development of new immunotherapies for the treatment of cancer using interleukin-2. Ann Surg 208:121–135
80. Grimm EA, Rosenberg SA (1983) The human lymphokine-activated killer cell phenomenon. In: Pick E, Candy M (eds) Lymphokines, vol 9. Academic, New York, pp 279–311
81. Rosenberg SA, Grimm EA, McGrogan M, Doyle M, Kawasaki E, Hoths K, Mark DF (1984) Biological activity of recombinant human interleukin-2 produced in Escherichia coli. Science 223:1412
82. Lotze MT, Franan LT, Ettinghausen SE et al (1985) In vivo adminstration of purified human interleukin-2. Half life and immunologic effects of the Jurkat line derived interleukin-2. J Immunol 134:157–166
83. Lotze MT, Matory EYL, Ettinghausen SE et al (1985) In vivo administration of purified human interleukin-2. Half-life, immunological effects and the expansion of peripheral lymphoid cells in vivo with recombinant IL-2. J Immunol 135:2865–2875
84. Matory YL, Chang AE, Brazil EH, Hyatt CL, Rosenberg SA (1985) Toxicity of recombinant human interleukin-2 in rats following intervenous infusion. J Biol Respir Mod 4:377
85. Teitelman R (1989) Gene dreams: Wall Street, academia and the rise of biotechnology. Basic Books, New York, pp 188–190
86. Rosenberg, SA, Lotze MA, Leitman S et al (1985) Observations on the systemic administration of autologous lymphokine-activated killer cells and recombinant interleukin-2 to patients with metastatic cancer. N Engl J Med 313:1485–1492
87. Editorial. Cancer patience (1985) The Wall Street Journal, 16–12–1985
88. De Vita V (1985) The search for a cure. Newsweek, December 16

89. Patterson J (1987) The dread disease: cancer and the modern American culture. Harvard University Press, Cambridge, pp 295–296
90. Durant JR (1987) Immunotherapy of cancer: the end of the beginning? N Engl J Med 316:939–940
91. Oldham RK (1987) Patient founded cancer research. N Engl J Med 316:46–47
92. West WH, Tauer KW, Yarnelli JR et al (1987) Constant infusion recombinant interleukin-2 in adoptive immunotherapy of advanced cancer. N Engl J Med 316:898–905
93. Lotze MT, Chang AE, Seipp CA et al (1986) High doses of recombinant interleukin-2 in the treatment of patients with disseminated cancer. JAMA 256:3117–3124
94. Moertel CG (1986) On lymphokines, cytokines and breakthroughs. JAMA 256:3141
95. Klausner A (1986) Clinicians question adoptive immunotherapy. Biotechnology 4:1044
96. Wang JCL, Walle A, Novogrodzky A et al (1989) A phase II clinical trial of adoptive immunotherapy for advanced renal carcinoma using mitogen-activated autologous lymphocytes and continuous-infusion interleukin-2. J Clin Oncol 7:1885–1891
97. Dutcher JP, Creekmore S, Weiss GR et al (1989) A phase II study of Interleukin-2 and lymphocyte activated killer cells in patients with metastatic malignant melanoma. J Clin Oncol 7:477–485
98. Clark JW, Smith JW, Steis RG et al (1990) Interleukin-2 and lymphocyte activated killer cell therapy: analysis of a bolus interleukin-2 and a continuous infusion interleukin-2 regimens. Cancer Res 50:7343–7350
99. Denicoff KR, Rubinoff DD, Papa MZ et al (1987) The neuropsychiatric effects of treatment with interleukin-2 and lymphokine-activated cells. Ann Intern Med 107:293–300
100. Jackson BS, Strautman J, Fredrickson K, Strauman TJ (1991) Long term biopsychosocial effects of interleukin-2 therapy. Oncol News Forum 18(4):693–690
101. Tatcher N, Dazzi H, Johnson RJ et al (1989) Recombinant interleukin-2 given intrasplenically and intravenously for advanced malignant melanoma. Br J Cancer 60:770–774
102. Negrier S, Philip T, Stoter G et al (1989) Interleukin-2 with or without LAK cells in metastatic renal cell carcinoma. A report of a European multicenter study. Eur J Clin Oncol 25 [Suppl 3]:21–28
103. Parkinson DR, Abrams JS, Wiernik PH et al (1990) Interleukin-2 therapy in patients with metastatic malignant myeloma. A phase II study. J Clin Oncol 8:1650–1656

104. Gosh AK, Dazzi HN, Tatcher N, Moore M (1989) Lack of correlation between peripheral blood lymphokine-activated killer (LAK) cell function and clinical response in patients with advanced malignant melanoma receiving recombinant interleukin-2. Int J Cancer 43:410–414

105. Favrot MC, Combaret V, Negrier S et al (1990) Functional and immunophenotypic modifications induced by interleukin-2 did not predict response to therapy in patients with renal cell carcinoma. J Biol Respir Mod 9:167–177

106. Sosman JA, Hank JA, Moore KH et al (1991) Prolonged interleukin-2 (IL-2) treatment can augment immune activation without enhancing antitumor activity in renal cell carcinoma. Cancer Invest 9:35–48

107. Smith KA (1989) Interleukin futures. Biotechnology 7:661–667

108. De Fano C (1912) A cytological analysis of the reaction in animals to implanted carcinomata. Sci Rep Imp Cancer Res Fund 5:57–78

109. Yron I, Spiess TA, Rosenberg SA (1980) In vitro growth of murine T cells: V. the isolation and growth of lymphoid cells infiltrating syngeneic solid tumors. J Immunol 125:238–245

110. Muul LM, Spiess PJ, Director EP, Rosenberg SA (1986) Identification of specific cytolytic immune responses against autologous tumor in humans bearing malignant melanoma. J Immunol 138:989–995

111. Rosenberg SA, Spiess P, Lafranière R (1986) A new approach to the adoptive immunotherapy of cancer with tumor infiltrating lymphocytes. Science 223:1318–1321

112. Rosenberg SA, Packard BS, Aebersold PA et al (1988) The use of tumor-infiltrating lymphocytes and interleukin-2 in the immunotherapy of patients with metastatic melanoma. New Engl J Med 319:1676–1680

113. Lotze MT, Custer MC, Bolton ES et al (1990) Mechanisms of immunologic antitumor therapy. Lessons from the laboratory and clinical applications. Hum Immunol 28:198–207

114. Rosenberg SA, Abersold P, Cornetta K et al (1990) Gene transfer into humans. Immunotherapy of patients with advanced melanoma using tumor-inflitrating lymphocytes modified by retroviral gene transduction. N Engl J Med 323:570–578

115. Cournoyer D, Caskey CT (1990) Gene transfer into humans: a first step. N Engl J Med 323:601–603

116. Mitchell MS, Kempf MA, Harel W et al (1989) Low dose cyclophospamide and low dose interleukin-2 for malignant melanoma. Bull N Y Acad Med 65:128–144

117. Marumo K, Muraki J, Ueno M et al (1989) Immunologic study of human recombinant interleukin-2 (low dose) in patients with advanced renal cell carcinoma. Urology 33:219–225

118. Atzpodien J, Körfer A, Evers P et al (1990) Low dose subcutaneous re-
     combinant interleukin-2 in advanced human malignancy: a phase II out-
     patient study. Mol Biother 2:18
119. Stein RC, Malkovska S, Morgan S et al (1991) The clinical effects of pro-
     longed treatment of patients with advanced cancer with low-dose subcuta-
     neous interleukin-2. Br J Cancer 63:275
120. Editorial (1989) Interleukin-2: sunrise for immunotherapy? Lancet i:308
121. Shiloni E, Pouillart P, Janssens J et al (1989) Sequential dacarbazine che-
     motherapy followed by recombinant interleukin-2 in metastatic myeloma.
     Eur J Cancer Clin Oncol 25 [Suppl 3]:45–49
122. Dilman RO, Oldham RK, Barth NM et al (1990) Recombinant inter-
     leukin-2 and adoptive immunotherapy alternated with decarbazine ther-
     apy in melanoma. A National Biotherapy Study Group Trial. J Natl Can-
     cer Inst 82:1345–1349
123. Tatcher N, Dazzi H, Mellor M et al (1990) Recombinant interleukin-2
     with flavone acetic acid in advanced malignant melanoma. A phase II
     study. Br J Cancer 61:618–621
124. Bergmann L (1989) Malignant melanoma: prognosis and actual treat-
     ment. Strategies with chemotherapy and biological response modifiers.
     Eur J Cancer Clin Oncol 25 [Suppl 3]:31–36
125. Parkinson DR, Talpaz M, Lee KH et al (1989) Interleukin-2 alone and in
     combination with other cytokines in melanoma: the investigational ap-
     proach at the University of Texas M.D. Anderson Cancer Center. Cancer
     Treat Rev [Suppl A]:39–48
126. Cicardelle T, Smith K (1989) Interleukin-2: prototype for the new genera-
     tion of immunoactive pharmaceuticals. Trends Pharmacol Sci 19:239–242
127. Hodgstone J (1991) Data directed drug design. Biotechnology 9:19–21
128. Burns R, Rifkin G (1990) Companies targeting drug delivery. Biotechnol-
     ogy 8:513–522
129. Gillis S, Williams DE (1998) Cytokine therapy: lessons learned and fu-
     ture challenges. Curr Opin Immunol 10(5):501–504
130. Kedar E, Klein E (1992) Cancer immunotherapy: are the results discour-
     aging? Can they be improved? Adv Cancer Res 59:245–322
131. Alexandroff AB, Robins RA, Murray A, James K (1998) Tumor immu-
     nology. False hopes-new horizons. Immunol Today 19(6):247–250
132. Brown RE (1973) Rockefeller medicine men: medicine and capitalism in
     America. The University of California Press, Berkeley and Los Angeles
133. Moulin AM, Löwy I (1983) La double nature de l'immunologie: L'his-
     toire de la transplantation rénale. Fund Sci 4:201–208
134. Blume S (1992) Insight and industry. MIT Press, Cambridge
135. Kenney M (1986) Biotechnology. The University-Industrial Complex.
     Yale University Press, New Haven

136. Bud R (1993) The uses of life. Cambridge University Press, Cambridge
137. Swann JP (1988) Academic scientists and the pharmaceutical industry. Cooperative research in twentieth century America. John Hopkins University Press, Baltimore
138. Bovet D (1988) Une chimie qui guérit: histoire de la découverte des sulphamides. Payot, Paris
139. Michich E, Fefer A (1983) Biological response modifiers: subcomittee report. NIH, Bethesda Ma

4 Interferon-α: From Pass Interference to Cytokine Networking

P. Fitzgerald-Bocarsly

4.1 Introduction

Interferon (IFN), and in specific, type I interferon, has the distinction of being the first cytokine, or as we now recognize, family of cytokines to be discovered. Interferon was first described by Isaacs and Lindenmann in 1957 [1] as a transferable factor obtained from virally infected chick amniochorion that "interfered" with subsequent viral replication in uninfected cultures. For the first two decades after its discovery, this role for interferon was explored from a multitude of angles, but the emphasis was on the antiviral roles of this substance. Only latter did the pleiotropic nature of the functions of IFN become appreciated, with discovery of its roles in cell growth regulation and immune modulation. This early stage of interferon research (and which, indeed continues today at a more mechanistic, molecular level) can be thought of as the "pass interference" stage of interferon research.

4.2 Pass Interference: A Model for Antiviral Action of Interferon

Pass interference is a term that describes an illegal action in the game of American football. American football is an unusual game that is the passion of many Americans. Indeed, certain football events such as the college bowl games around New Years and the Super Bowl, which is the championship game for professional football, have broad cultural appeal in America. A recent retrovirus conference held in Chicago, Illinois had the opening session and welcome reception on the night of the Super Bowl. These sessions were notably depleted of American participants, who presumably were in their hotel rooms or hotel bars huddled around the televisions, enjoying the game, food and camaraderie. The game of football is also interesting in that it really has very little to do with feet. Unlike its distant cousin, European football or soccer, in American football, the only time the ovoid ball ever touches a foot is during kickoffs, when control of the ball is changing to the opposing side (usually on fourth down) or when a field goal or extra point attempt is made. For the majority of the game, the ball is either carried up the field ("rushing") or passed to a receiver, who then runs up the field with the ball. In a passing play, the job of the defense is to prevent the ball from being caught by the receiver since a completed pass moves the "line of scrimmage" closer to the offensive goal. In pass interference, a defensive player "interferes" with the receiver before the ball is caught by the receiver (Fig. 1). In football, this action is illegal and results in a penalty. The analogy of pass interference can be used to describe the way in which interferon affects passing on of virus to uninfected cells (Fig.2). In this analogy, the quarterback is a cell infected with virus (the ball). The goal of virus is to be passed along to the uninfected cells (in this case, the pass receiver). Interferon, then is the defensive player who interferes with the pass reception. In pass interference, the defensive

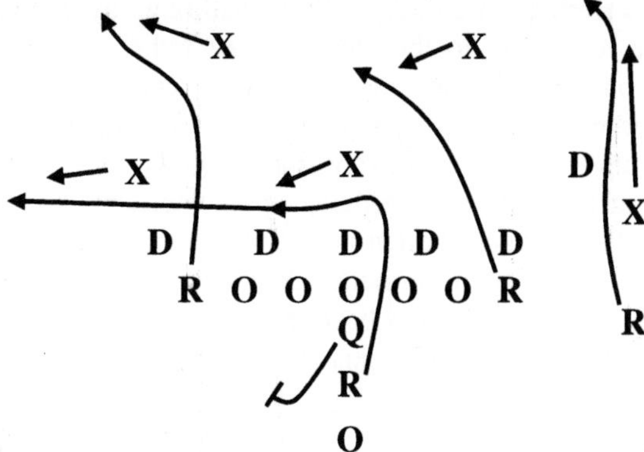

Fig. 1. Football: pass interference. In this "bootleg pass to left" play, the quarterback (*Q*) is attempting to pass the ball to the far left receiver (*R*). The defensive player (*X*), interferes with the receiver before he catches the ball. This play results in a penalty against the defense

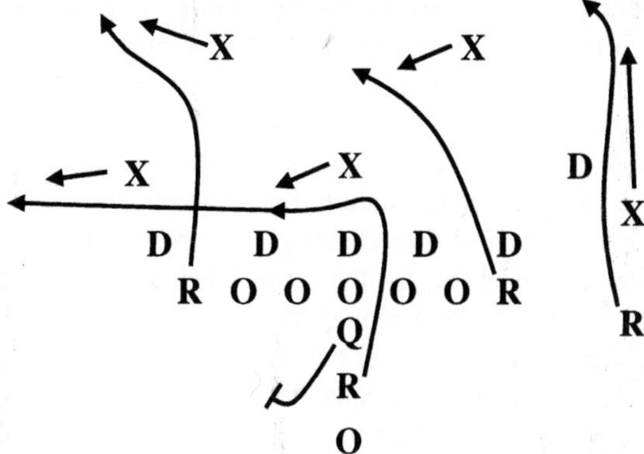

Fig. 2. Pass interference: the interferon analogy. In this "play", the infected cell (*Q*) is attempting to pass the virus onto the uninfected virus target (*R*). The far left interferon (*X*) renders the uninfected cell resistant to the infection, thereby interfering with the passing on of the virus. Note that there are other defensive players that can directly interact with the virus (i.e., antibody) or with virally-infected cells (e.g., natural killer cells and cytotoxic T cells). This "play" results in host protection from viral spread

player is interfering not directly with the ball (e.g., by intercepting the pass before it can get to the receiver), but rather with the receiver himself. In our analogy, then, the interferon reaches the uninfected cell (the receiver) before the ball, hampering the ability of the virus (the ball) to advance up the field, effectively shutting down viral replication by prevention of new infection. Although pass interference in football is technically illegal, it is still often successfully undertaken, i.e., by subtle interference. Certainly, in antiviral resistance the "pass interference" model is very important as has been recently demonstrated using mice that have had various components of the interferon signaling system "knocked-out" [2].

4.3 Host Responses to Virus: Beyond Pass Interference

The football analogy of viral resistance can be taken further than pass interference by interferon. In football, a team with just one defensive player would be rapidly overwhelmed. Viruses, as rapidly evolving organisms, often mutate to escape host defense mechanisms. An example of this is the appearance (or selection) of HIV variants that are resistant to the antiviral effects of interferon [3]. Therefore, just as there are other defensive players in the game of football, there must be other mechanisms of antiviral resistance. For example the ball (or, in our analogy, the virus) can be directly intercepted by the defense when it is in the air being passed (or, for the virus, when it is extracellular). An example of this mechanism would be neutralization of free virus by antibody. Alternatively, infected cells can be killed by cell-mediated effectors such as natural killer cells or cytotoxic T-cells. In the football analogy, this would be equivalent of sacking the quarterback or other offensive player who is in the possession of the ball. This "team" analogy can certainly be extended to the body's response to viral infection: for any given virus/host interaction, a certain component of the antiviral response is most critical; to successfully eradicate infection, however, it typically requires a number of different antiviral and immunological interactions. Interestingly, many of these other antiviral defense mechanisms do not function independently of interferon. Indeed, we now know that interferons, both type I and type II are important modulators of the immune response, as will be discussed below.

4.4 Interferon in Cancer Therapy

In addition to its antiviral and immunomodulatory roles, interferon is a potent inhibitor of cell growth. In the early years before the advent of recombinant interferon, extension of in vitro studies to human treatment was hampered by the difficulty in obtaining sufficient amounts of purified material. This was partially overcome by the important observation by Kari Cantell that human peripheral blood mononuclear cells could be induced to produce large amounts of interferon in response to stimulation with the parainfluenza virus, Sendai virus [4]. The massive cultures of human blood buffy coats with Sendai virus carried out by the Finnish group in conjunction with the Finnish Red Cross provided for the first time sufficient material that could be tested in humans. Although it is now recognized that the purity of the interferon obtained from these cultures was extremely low, there was much demand for this product for human studies, and some efficacy was shown for the interferon treatment of certain malignancies. Following the partial purification of natural leukocyte IFN, the cloning of IFN-α and -β were critical to the advancement of the field (reviewed in [5]).

Recombinant IFNs allowed for the first time large quantities of uniform material that could be tested in the clinic. Although the original hopes for IFN-α as the "magic bullet" cure for cancer proved to be overly optimistic, today recombinant IFN occupies important clinical niches for the treatment of specific malignancies and viral infections.

4.5 Interferon as Immunomodulator

Recognition of interferon as an immunomodulator came in the late 1970s, with observations that it was able to upregulate the activity of a newly described population of cells, the natural killer (NK) cells [6] (Fig. 3). NK cells are traditionally considered to be components of the innate, or first-line defense system since they do not require prior sensitization or establish immunological memory. These cells are active in the killing of virally infected target cells and certain tumor targets in vitro, and the absence of NK cells is associated with increased severity of certain viral infections such as those in the herpes virus family. Interferon was found to have the dual role of augmenting the cytolytic

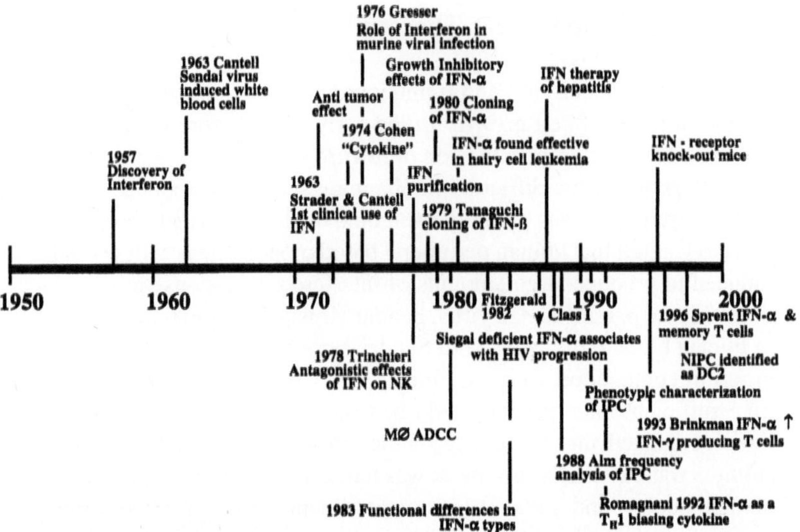

Fig. 3. An abbreviated history of interferon from 1957–1999. Some key events in interferon-α research. Note that immunological advances are shown below the *time line*. The author entered the interferon field in 1982

activity of the NK cells as well as protecting uninfected cells from lysis [7]. Thus, the model was proposed whereby interferon would be induced in viral infection, serving to upregulate the NK activity against virally infected cells while also protecting uninfected cells from bystander attack. IFN was also found to positively regulate the cytolytic activity of macrophages against intracellular parasites. This activity was most efficiently induced by type II IFN or IFN-γ (called "macrophage activating factor" in the original nomenclature), but could also be affected by the type I IFNs.

Based on these initial observations, for a number of years, IFN-α was viewed by the immunological community as primarily a component of the innate immune response. In fact, the innate response, and those who studied it were often considered to be outside of the mainstream of immunology, which focused on the adaptive immune response mediated by the antigen-specific effectors: antibody produced by B-cells and recognition of specific antigen by T-cell subsets. This compartmentalization of the immune system into distinct innate and adaptive responses, while providing a useful paradigm for antigen recognition, has proven to be incomplete. Over the last several years, there has been an increasing appreciation that there is not distinct compartmentalization of the immune responses; rather, the immune system is seen as a complex network, with a continuum of responses and multiple mechanisms of communication and control occurring between the various components [8,9]. Moreover, there are multiple levels of overlap of the innate and adaptive responses, with IFNs carrying out some of the important functions. An example of this is IFN-γ: this cytokine is produced not only in an antigen specific manner by T-helper cells (i.e., as part of the adaptive response), it is also produced by NK cells (considered to be components of the innate response). Moreover, this cytokine is recognized as going far beyond its originally defined role as macrophage activating factor (innate immunity) to being involved in directing T-helper 1 responses required for cell mediated immunity (adaptive immunity). Interleukin (IL)-12, a cytokine produced by a variety of antigen presenting cells, is also seen as a major link in the innate to adaptive continuum of responses: failure to produce IL-12 in the early, innate responses results in a failure to produce normal adaptive, and in particular, the T_H1 cytokine dependent cell-mediated immune responses.

4.6 IFN-α as a T_H1 Biasing Cytokine

IFN-α, like IFN-γ and IL-12, is now being appreciated for its function in bridging the innate to adaptive continuum. In humans, IFN-α is a T_H1-biasing cytokine; that is, it helps to bias towards cytokine production that favors cell-mediated immunity [10–12]. Cell-mediated immune functions are required for the elimination of most intracellular parasites and viruses. Since IFN-α is typically induced in response to such infectious agents, it makes sense that this would lead to appropriate protective immunity. IFN-α is known to enhance the expression of IFN-γ by T-helper cells [13], and in mice, it has been shown to be involved in the development and maintenance of memory cytotoxic T-cells [14]. Moreover, IFN-α is known to upregulate the expression of the receptor for IL-12, thus enhancing the T_H1 responses induced through this latter cytokine.

4.7 IFN-α Producing Cells in Human Peripheral Blood

Over the years, the majority of studies of the IFN-α producing cells in peripheral blood have been carried out using Sendai virus as the IFN inducer, the original virus used by Cantell for the production of the natural leukocyte product named Finnferon [4]. Indeed, Sendai virus is still the virus being used to produce the currently available leukocyte or lymphoblastoid products. An early study by Saksela et al. [15] convincingly demonstrated that the predominant cells that produce IFN-α in response to Sendai virus are monocytes. In fact, to this day, most textbooks and reviews will cite monocytes as the IFN producing cells in peripheral blood. The focus on Sendai virus-induced IFN-α, however, appears to have led researchers to a less prevalent pathway of IFN production. It is now clear that the monocyte is not the only IFN-α producing cell in human peripheral blood. Indeed, the high induction of IFN-α by Sendai virus in monocytes appears to be a rather unique phenomenon.

A number of groups, including our own, have studied the production of IFN-α in response to a variety of enveloped DNA and RNA viruses and found that for these other viruses, a non-monocytic cell was the primary producer of IFN-α (reviewed in [16]). This cell, which was initially called the "natural interferon-α producing cell" or "NIPC", is found at low frequency in peripheral blood, with functional NIPC representing approximately 0.1% of peripheral blood mononuclear cell (PBMC). Based on frequency analysis (as determined by ELISpot assay) and total IFN-α production, it is estimated that each NIPC can produce from 1-2 IU of IFN-α when stimulated by herpes simplex virus, which is equivalent to 3-10 pg IFN/cell! The NIPC lacks cell surface markers typical of the T, B, NK or monocyte lineages but does express HLA-DR and CD4. This phenotype is identical to that of the circulating immature dendritic cells, and enrichment studies and intracellular flow cytometric studies for IFN-α producing cells have indicated that these cells are the NIPCs [17–20]. Most recently, it has been recognized that there are at least two circulating dendritic cell populations in the peripheral blood CD123 [21,22]: one expresses low levels of receptors for granulocyte macrophage–colony-stimulating factor (GM-CSF), is CD11c- but expresses high levels of the receptor for IL-3. The second population is GM-CSF receptor high, CD11c+, and CD123 low. The

NIPCs have recently been shown to belong to the former cell type, with virtually all of the NIPCs expressing CD123 but not CD11c, as determined by intracellular flow cytometry for IFN-α expression combined with cell surface phenotype analysis or cell sorting (Fitzgerald-Bocarsly et al., in preparation).

The CD11c- cell has been recently termed a precursor to the type 2 dendritic cell (pDC2) [23], a cell that is known to travel from the blood to the T-cell areas of lymphoid tissue, where it resides near the high endothelial venules. In contrast, the CD11c+ (pDC1) cells home to germinal centers. The localization of the pDC2 in lymphoid tissue is intriguing. The presence of these cells in lymphoid tissues has long been noted by pathologists. The cells were often referred to as "plasmacytoid T-cells" or "plasmacytoid monocytes" because of their morphological appearance [24]. On transmission electron microscopy, the cells have abundant cytoplasm with well-developed arrays of endoplasmic reticulum, being clearly prepared to produce large quantities of protein for secretion. IFN-α is the first known secretory product of these cells. We hypothesize that the NIPCs pick up virus in the periphery, then carry it to the lymphoid tissue where the cells will release IFN-α. The NIPCs then mature into antigen presenting DC2 cells, facilitating a T_H1 response that will enhance cell-mediated immunity to clear the viral infection. This hypothesis is supported by our recent observations that IFN-α upregulates costimulatory molecule expression on the NIPCs, and also serves as a survival factor for these cells (unpublished observations).

Major roles for the NIPCs in antiviral resistance are supported by their deficient function in patients with the acquired immune deficiency syndrome [25,26]. Deficient IFN-α production in response to herpes simplex virus (HSV), the prototypical inducer of IFN production by NIPCs occurs relatively early in disease and is predictive of subsequent opportunistic infections and mortality. In contrast, the IFN-α production by monocytes in response to Sendai virus was found to occur only very late in disease and did not correlate with disease progression [27,28]. The mechanisms for dysfunction of the NIPC in HIV infection is currently unknown; however, functional studies of the frequency of the NIPCs as determined by ELISpot indicate that both decreased numbers of cells secreting IFN-α as well as decreased IFN-α production on a per cell basis contribute to the deficiency. The recent definition of the phenotype of the NIPCs (CD11c-, CD123+, HLA-DR+, lineage negative) provides for the first time a mechanism for tracking the cells phenotypically in peripheral blood, and will enable us to determine how their numbers are affected in HIV infection and other diseases.

4.8 Differences Between IFN Production by NIPCs and Monocytes

As described above, the commercial leukocyte and lymphoblastoid interferon preparations have all been produced using Sendai virus as the stimulating virus. The reasons for this are probably largely historical: Cantell found that Sendai virus, and in particular, specific strains of the virus were very high inducers of IFN-α production. It is interesting to note that some other viruses such as herpes simplex virus can also induce high levels of IFN production from freshly isolated PBMC, but not from overnight cultured PBMC. We now know that the NIPCs (and pDC2) are highly culture labile, rapidly dying, probably by apoptosis. Thus, the lengthy times from obtaining buffy coats of blood to culturing with virus undoubtedly selects against the viability of NIPCs.

In addition to time issues in blood processing, some of the techniques used for producing leukocyte IFN involve lysing of red blood cells with ammonium chloride. Ammonium chloride is a weak base that, in addition to lysing RBCs, also neutralizes the acidic compartments of cells. Ammonium chloride and other inhibitors of endosomal acidification or endocytosis are inhibitors of IFN production by NIPCs [29]. These treatments, however, have no effect on production of IFN-α by monocytes [30]. One possible explanation for these differences is the mechanisms by which Sendai virus vs. other cells enter into cells. Sendai virus is able to fuse with the cellular envelope at neutral pH, whereas some other viruses require the acidic pH of the endosome for viral envelope fusion.

In a study where we evaluated the frequency of interferon producing cells in response to a variety of different DNA and RNA viruses, we observed that the majority of these viruses induced IFN in a low frequency population (0.05-0.1%, as determined by ELISpot assay), and that this IFN production was sensitive to the lysosomotropic drug, chloroquine [30]. Only Sendai virus induced IFN was found to be

Table 1. Summary of HSV and Sendai-responsive IPC

	HSV-responsive	Sendai-responsive	Reference
Frequency (by ELISPOT)	0.1%	0.5%–2%	[30]
IFN produced/IPC	3–10 pg	0.3%–0.5 pg	[26]
Inhibition by IL-4	No	Yes	[34]
Chloroquine, bafilomycin, ceramide sensitivity	Yes	No	[29,30], our unpublished
Blocked by anti-MR antibody	Yes	No	[17]
Sensitivity to overnight culture	Yes	No	[16]

HSV, herpes simplex virus; IFN, interferon; IPC, interferon-producing cell

chloroquine independent, and to occur at high frequency, with as many as 1%-2% of the PBMC responding with IFN-α production. A notable exception was the interferon production in response to M-tropic strains of HIV-1: HIV-1$_{BAL}$ and -HIV-1$_{ADA}$. The interferon response to each of these viruses included a distinct monocytic compartment, with both CD123+ DC and CD14+ monocytes producing IFN-α in response to this virus (Dahmani et al., in preparation).

There are a number of other differences between IFN induction by Sendai virus and HSV (Table 1). Notably, the monocytes, while representing more frequent IFN producing cells, produce approximately 10-fold less IFN-α on a per cell basis than do the NIPCs [30]. Whereas UV-inactivated virus induces IFN as well as live virus, UV-inactivated Sendai virus does not induce IFN. This result suggests that active viral infection is required for induction of IFN in monocytes. Finally, we have recently defined the mannose receptor (MR), which is found on dendritic cells and macrophages, but not monocytes as being involved in the production of IFN-α by NIPC [17]. According to our model, the mannose receptor, which recognizes repeating carbohydrate units, serves as a pattern recognition receptor for the uptake of a variety of enveloped viruses into the endosomal compartments of DC. Antibody to the mannose receptor blocks IFN-α production in response to HSV, VSV and HIV but does not interfere with IFN production in response to Sendai virus.

The dissimilarities between IFN production by monocytes and NIPCs raises the important philosophical question about the natural IFN products currently available. These interferons are derived by stimulating PBMCs or whole blood leukocytes with Sendai virus or by stimulating the lymphoblastoid cell line Namalwa with this virus. Given the heterogeneity of interferon producing cells and the multiple IFN-α genes that are expressed (to be discussed below), it is possible that these commercial preparations are missing some of the functions that might be derived from dendritic cell interferon.

4.9 Differential Induction of IFN

Research on IFN-α is complicated not only by the different cells capable of producing this protein, but also by the fact that IFN-α is really a family of related proteins, all of which bind to the same receptor. In fact, IFN-β and IFN-ω also bind to the same receptor. In humans, although there are single genes encoding IFN-β and IFN-γ, there are multiple IFN-α genes and pseudogenes, and leukocyte IFN-α preparations contain a mixture of subtypes (reviewed in [5]). Although the IFN-α subtypes all bind to a common receptor, it is clear that they do not have equal functions. An example of this is IFN-α7. Although this subtype has good antiviral activity, it fails to augment NK cell activity [31]. Among the subtypes, IFN-α8 is known to have the most potent antiviral activity, at least against some viruses such as HIV-1.

The existence of so many subtypes of IFN-α has intrigued scientists. In his recent memoir, Kari Cantell, commenting on the fact that Finnferon contains 13 different alpha interferons as well as omega interferon, wrote

The key question is why human cells produce so many different alpha interferons. Do they represent just evolutionary accidents, a profligate waste of resource on the part of the Nature? Would a single alpha interferon be capable of performing the same functions as all the other alpha interferons? Or is it that each different alpha interferon has its own, presently unknown, special function in the human body? [32].

Although the general consensus is that the different IFN-α subtypes are maintained in the genome for some purpose, and do not represent functional redundancy, the overall significance of the multiple genes and their products raises both scientific questions and practical concerns about our current therapeutic use of IFN-α.

The multiplicity of interferon-α genes has also led to some difficulties in studying their function in vivo. A popular way of studying the effects of individual cytokines has been to knock-out the cytokine gene in embryonic murine stem cells, then observe the phenotype in the resultant animal. For cytokines encoded by a single gene, such studies have been relatively straightforward, although the interpretation of the results of such experiments are not always clearcut. For example, cytokine knockout mice have taught us that there is a great deal of partial redundancy in cytokines, with other cytokines sometimes filling in, at least partially, for the absent cytokine. Direct knockout experiments of IFN-α have been precluded by the multiple IFN-α genes present in the murine genome. Instead, Aguet and his colleagues have created mice unable to respond to type I IFNs by knocking out the IFN-α/β receptor [2]. Other studies have focused on knocking-out components of the IFN-α signaling pathway such as the IRFs and STAT proteins.

Among the pressing questions regarding the IFN-α subtypes are: (1) How do the different subtypes transduce differential signals through a common receptor? (2) What controls the relative ratios of the IFN-α gene products? Is the ratio of the IFN-α subtypes controlled by the nature of the inducer, the cell type producing the IFN-α (e.g., NIPC vs. monocyte), or might the general cytokine milieu be contributing to the mix? (3) Does IFN-α production that leads to T_H1 biasing differ from IFN-α that preferentially induces an antiviral state? (4) How is the apparent potentiation of some IFN-α subtypes for each other accomplished at the molecular level?

In the realm of clinical practice, the existence of multiple subtypes of IFN-α raises practical questions. Both of the dominant recombinant IFN-α preparations on the market, Roferon (produced by Roche) and Intron A (produced by Schering) are clones of IFN-α2. These two recombinant proteins differ by only a single amino acid. Each of these was cloned as prominent IFN-α message that appeared in cultures of Sendai virus-stimulated peripheral blood mononuclear cells. What might the clinical effects be of some other recombinant IFN-α subtypes be? For example, our evidence suggests that the mix of IFN message differs between HSV and Sendai virus-stimulated PBMCs. The commercially available "natural" IFN-α products from lymphoblastoid cells or leukocyte cultures stimulated with Sendai virus contain mixtures of IFN subtypes. These mixtures have at least one advantage over the recombinant products in that they can be tolerated in patients over longer periods of time without the development of neutralizing antibodies. However, given that different populations of cells produce IFN-α which may have differential effects, one wonders whether Sendai induced IFN-α is necessarily the "best" IFN-α preparation for treatment of a range of diseases, from viral infections such as hepatitis C to tumors, or in replacement immunotherapy. Might the best mix of IFNs vary for the type of disease being treated?

From a practical point of view, the costs of answering these questions and of potentially bringing complex mixtures of IFNs to the clinic are probably insurmountable. It is certainly true that many promising biological drugs never make it to the clinic because the preliminary studies were not carried out in exactly the correct way to prevent toxicity and/or maximize efficacy. For IFN-α and the complexity of its subtypes, these issues become even more difficult. That the first cloned IFN-α has been efficacious in treatment of a number of viral and neoplastic conditions is quite remarkable, and certainly raises the question of how much more effective appropriately designed cocktails of the IFNs might be. Clearly, preclinical studies of the interferon subsets and the cells that produce them are warranted. Whether such knowledge can be successfully translated to the clinic in an economically viable manner remains to be seen.

4.10 Interferon Networking

Interferons, both type I and type II are members of a group of proteins that communicate between cells of the immune system, called cytokines. Unlike hormones, which typically work at sites distal to the site of production, cytokines typically exert their functions over relatively short distances, functioning in an autocrine or paracrine manner. Two general properties of cytokines are that they are pleiotropic, i.e. they have multiple functions, and their functions tend to be at least partially redundant. IFN-α certainly meets these latter two criteria: they have multiple functions, for example, induction of antiviral defenses, anti-cell cycling mechanisms, immunomodulatory functions. They are also express redundant functions with some other cytokines. One example of this is that natural killer cell activity can be upregulated by both IFNs and IL-12.

Scientists who study cytokines and/or cells in the immune system enjoy making complicated charts of how their cytokines and cells fit into the overall scheme or "network" of the immune response. The cytokine network is growing increasingly complex as more and more cytokine pathways and interactions, both positive and negative, among cytokines are described. An analogy to the cytokine network is the complex freeway system in Los Angeles, California. Most visitors to Los Angeles will come away with a vision of the seemingly incomprehensible interchanges that interconnect a series of freeways throughout greater Los Angeles. To get from point A to point B, (for example, from the San Fernando Valley to Pasadena), there are several distinct freeway paths one can follow. Listening to news radio stations that broadcast freeway traffic reports every 6 min or so is critical to the driver attempting to navigate the freeway system during rush hour. An accident or traffic back-up in one portion of the freeway system can significantly affect forward progress not only on that freeway, but at interchange points, and subsequently on other freeways as well. The freeway analogy works nicely for the interferon pathways and how they interact with other cytokines.

Up until a few years ago, IFN-α and the cells that produce it were conspicuously absent from most hematopoietic and cytokine networking charts. Our current understanding of the cells that produce IFN and their relationship to the immune system are rapidly changing that picture. We have recently described the negative regulation of virally-induced IFN-α by the T_H2 cytokine IL-10 [33]. Addition of IL-10 to PBMC cultures stimulated with either HSV or Sendai virus results in down-regulation of the IFN-α production at the transcriptional level. Moreover, neutralization of endogenously produced IL-10 by the addition of antibody to IL-10 results in an increase in both the frequency of the IFN-α producing cells as well as the total IFN produced by these cells. These latter results would suggest that IL-10 must be endogenously produced during viral / PBMC culture. Indeed, IL-10 is produced by the PBMCs in response to a number of viruses. We hypothesize that IL-10 functions, in part, to prevent the overproduction of IFN-α. Such overproduction of IFN-α could be detrimental to the host, leading to significant pathology and autoimmunity as is frequently seen in patients receiving IFN therapy. Perhaps this inhibition of IFN-α production also helps control the induction of T_H1 responses. Certainly, IL-10 is a well-known inhibitor of the T_H1 response, and is associated with T_H2-mediated humoral responses. The virus-induced IL-10 is produced by monocytes (Payvandi et al., submitted for publication), and therefore functions in a paracrine manner when it inhibits dendric cell produced IFN-α. However, it also inhibits Sendai induced IFN-α by monocytes; in this instance, we don't know whether the same cells are producing both IFN-α and IL-10. Interestingly, IL-4, another T_H2 cytokine, inhibits monocyte but not dendritic cell produced IFN-α, and is thus only partially redundant with IL-10 in its regulation of IFN-α production. Although we only are beginning to understand the roles for IFN-α in development of T_H1 immunity, it is clear that this cytokine clearly belongs to the complex cytokine network, and that the effects of manipulation of IFN-α or IL-10 on the overall homeostasis must be taken into account.

4.11 The Future of Interferon

One of the difficulties facing the scientist who studies IFN-α is the problem of getting the research funded. In American medical schools and biomedical research institutions, the main source of external research funding has historically been the National Institutes of Health. Investigator-initiated grants, the "RO1" grants, are the mainstay of these investigators. Grants are evaluated by groups of reviewers called "Study Sections" with primary and secondary, and sometimes tertiary reviewers assigned to each proposal. The results of the review are communicated to the investigator by means of the summary statement, which is still popularly called the "pink sheet", a name that reflects the color of the paper the reviews used to be printed upon. Just as clothing, language and even morality are trendy, biomedical research areas are also trendy. Go to any meeting, and the latest "trend" in science will be reflected in the choice of symposium topics and speakers. Interferon, and in particular IFN-α is, in many respects, "old news". As the earliest cytokines described, the type I interferons have been extensively studied over the last 40 years, but often fail to capture the imagination of today's scientist. One of my "favorite" comments from a set of pink sheets for a renewal proposal I submitted several years ago was the statement "New cytokines and pathways compete for attention, such as the recent fame of IL-12." This comment was not meant to be derogatory towards the research in my proposal, but rather to point out that in a time of tight money, there were perhaps other "sexier" topics to capture the imagination of the study section. This statement was written in 1994, and perhaps "IL-12" would today be replaced by IL-18 or some of the chemokines and their receptors, the latter having gained recent fame for their role in HIV infection. Thus, one of the most important roles of the interferonologist is to keep the importance of this family of cytokines in a prominent position: only the surface of the biology and potential applications for interferons and their signaling proteins has been grazed.

For the inteferonologist in a private pharmaceutical company, the problem is even more difficult. Here the issues are not only what constitutes good science, but also what products can be successfully navigated through the myriad of steps before they actually reach the clinic. Here again, new cytokines and pathways compete for attention. While it is interesting to consider the potential of different interferon subtypes, either as natural or recombinant products as differential therapeutics, the economics of testing such drugs or combinations of drugs may make the issues moot. Clearly, if such combinations are ever to be tested in patients, then the basic scientists will have to make compelling arguments for the reasonableness of these approaches. Additionally, the use of interferon as an adjuvant (as opposed to sole) therapy is only beginning to be approached.

Additionally, although much has been learned about interferons and other cytokines, there is still much to learn about how to efficiently deliver the cytokines to patients in a beneficial manner. One of the major concerns in interferon therapy has been the associated toxicity of the therapy and the subsequent development of neutralizing anti-IFN antibodies and autoimmunity. Other cytokines such as IL-2, TNF-α and IL-12 have also demonstrated significant toxicity in animal and patient trials, and have even lead to death. A major problem with current cytokine delivery methodology is the systemic administration of the cytokines. Cytokines typically act locally, achieving locally high concentrations, but do not usually establish high systemic concentrations. Delivery of the cytokine locally would be predicted to generate less toxicity and unpleasant side effects. Preliminary data suggest that delivery of IFN-α or IL-12 locally with tumor vaccines can augment TH1 responses and protective immunity. Such delivery of cytokines might be achieved in vaccines by incorporating either the protein or the genes for the cytokines into the vaccine formulation.

Certainly our understanding of interferon and its role in host resistance has exponentially increased over the years. Our understanding of IFN-α has moved well beyond pass interference to the realm of cytokine networking of the complex IFN signaling pathways. As we approach the next millenium, there is still much to be learned about interferon and how to harness it more effectively for the clinic. The long, productive history of interferon research provides an excellent framework on which the current and future generations of interferonologists can build.

References

1. Isaacs A, Lindenmann J (1957) Virus interference. 1. The interferon. Proc R Soc Biol 147:258
2. Muller U, Steinhoff U, Reis LFL, Hemmi S, Pavlovic J, Zinkernagel RM, Aguet M (1994) Functional role of type I and type II interferons in antiviral defense. Science 264:1918–1921
3. Edlin B, St Clair M, Pitha P, Whaling S, King D, Bitran J, Weinstein R (1992) In vitro resistance to zidovudine and alpha interferon in HIV-1 isolates from patients: correlations with treatment duration and response. Ann Intern Med 117:457
4. Cantell K, Hirvonen S, Kauppinen H-L, Myllyla B (1981) Production of interferon in human leukocytes from normal donors with the use of Sendai virus. Methods Enzymol 78:29–38
5. Pestka S, Langer JA, Zoon KC, Samuel CE (1987) Interferons and their actions. Annu Rev Biochem 56:727–777
6. Trinchieri G, Santoli D, Koprowski H (1978) Spontaneous cell-mediated cytotoxicity in humans: role of interferon and immunoglobulins. J Immunol 120:1849–1855
7. Trinchieri G, Santoli D, Granato D, Perussia B (1981) Antagonistic effects of interferons on the cytotoxicity mediated by natural killer cells. Fed Proc 40:2705–2710
8. Medzhitov R, Janeway C (1998) An ancient system of host defense. Curr Opin Immunol 10:12–15
9. Medzhitov R, Janeway CA (1997) Innate immunity: impact on the adaptive immune response. Curr Opin Immunol 9:4–9

10. Parronchi P, De Carli M, Menetti R, Simonelli C, Sampognaro S, Piccinni M-P, Macchia D, Maggi E, Del Prete G, Romagnani S (1992) IL-4 and IFN (α and γ) exert opposite regulatory effects on the development of cytolytic potential by Th1 or Th2 human T cell clones. J Immunol 149:2977–2983
11. Del Prete G, Maggi E, Romagnani S (1994) Biology of disease, human Th1 and Th2 cells: functional properties, mechanisms of regulation, and role in disease. Lab Invest 70:299–306
12. Romagnani S (1992) Induction of TH1 and TH2 responses: a key for the natural immune response? Immunol Today 13:379–381
13. Brinkmann V, Geiger T, Alkan S, Heusser, CH (1993) Interferon α increases the frequency of interferon γ-producing human CD4+ T-cell. J Exp Med 178:1655–1663
14. Tough DF, Borrow P, Sprent J (1996) Induction of bystander T-cell proliferation by viruses and type I interferon in vivo. Science 272:1947–1950
15. Saksela E, Virtanen I, Hovi T, Secher DS, Cantell K (1984) Monocyte is the main producer of human leukocyte alpha interferons following Sendai virus induction. Prog Med Virol 30:78–86
16. Fitzgerald-Bocarsly P (1993) Human natural interferon-α producing cells. Pharmacol Ther 60:39–62
17. Milone M, Fitzgerald-Bocarsly P (1998) The mannose receptor mediates induction of IFN-α in peripheral blood dendritic cells by enveloped RNA and DNA viruses. J Immunol 161:2391–2399
18. Feldman M, Fitzgerald-Bocarsly P (1990) Sequential enrichment and immunocytochemical visualization of human interferon-α producing cells. J Interferon Res 10:435–446
19. Svensson H, Johannisson A, Nikkila T, Alm GV, Cederblad B (1996) The cell surface phenotype of human natural interferon-a producing cells as determined by flow cytometry. Scand J Immunol 44:164–172
20. Ferbas JJ, Toso JF, Logar AJ, Navratil JS, Rinaldo CR (1994) CD4[+] blood dendritic cells are potent producers of IFN-α in response to in vitro HIV-1 infection. J Immunol 152:4649–4662
21. Olweus J, BitMansour A, Warnke R, Thompson P, Carballido J, Picker L, Lund-Johansen F (1997) Dendritic cell ontogeny: a human dendritic cell lineage of myeloid origin. Proc Natl Acad Sci USA 94:12551–12556
22. O'Doherty U, Peng M, Gezelter S, Swiggard WJ, Betjes M, Bhardwaj N, Steinman RM (1994) Human blood contains two subsets of dendritic cells, one immunologically mature and the other immature. Immunology 82:487–493
23. Rissoan M-C, Sournelis V, Kadowaki N, Grourard G, Briere F, de Waal Melefyt R, Liu Y-J (1999) Reciprocal control of T-helper cell and dendritic cell differentiation. Science 283:1183–1186

24. Grouard G, Rissoan M, Filguiera L, Durand I, Banchereau J, Liu J (1997) The enigmatic plasmacytoid T-cells develop into dendritic cells with interleukin-3 and CD40 ligand. J Exp Med 185:1101–1111
25. Siegal FP, Lopez C, Fitzgerald PA, Shah K, Baron P, Leiderman IZ, Imperato D, Landesman S (1986) Opportunistic infections in acquired immune deficiency syndrome result from synergistic defects of both natural and adaptive components of cellular immunity. J Clin Invest 78:115–123
26. Howell D, Feldman S, Kloser P, Fitzgerald-Bocarsly P (1994) Decreased frequency of natural interferon producing cells in peripheral blood of patients with the acquired immune deficiency syndrome. Clin Immunol Immunopathol 71:223–230
27. Feldman SB, Milone MC, Kloser P, Fitzgerald-Bocarsly P (1995) Functional deficiencies in two distinct IFN-α producing cell populations in PBMC from human immunodeficiency virus seropositive patients. Leuk Biol 57:214–220
28. Ferbas J, Navratil J, Logar A, Rinaldo C (1995) Selective decrease in human immunodeficiency virus type 1 (HIV-1)-induced alpha interferon production by peripheral blood mononuclear cells during HIV-1 infection. Clin Diag Lab Immunol 2:138–142
29. Lebon P (1985) Inhibition of herpes simplex virus type-1 induced interferon synthesis by monoclonal antibodies against viral glycoprotein D and by lysosomotropic drugs. J Gen Virol 66:2781–2786
30. Feldman SB, Ferraro M, Zheng H-M, Patel N, Gould-Fogerite S, Fitzgerald-Bocarsly P (1994) Viral induction of low frequency interferon-α producing cells. Virology 204:1–7
31. Ortaldo JR, Herberman RB, Harvey C, Osheroff P, Pan, YCE, Kelder B, Pestka S (1984) A species of human α interferon that lacks the ability to boost human natural killer cell activity. Proc Natl Acad Sci USA 81:4926–4929
32. Cantell C (1998) The story of interferon: the ups and downs in the life of a scientist. World Scientific, Singapore
33. Payvandi F, Amrute S, Fitzgerald-Bocarsly P (1998) Exogenous and endogenous IL-10 regulate interferon-alpha production by PBMC in response to viral stimulatoin. J Immunol 160:5861–5868
34. Gobl A, Alm G (1992) Interleukin-4 down-regulates Sendai virus-induced production of interferon-α and -β in human peripheral blood monocytes in vitro. Scand J Immunol 35:167–175

5 A Biosemiotic View of Interferon: Toward a Biology of Really Living Organisms

Y. Kawade

5.1 Introduction

This symposium has the title "The dawn of recombinant protein drugs". But what kind of a dawn is it? Perhaps many of the participants will have an image of a dawn to a bright future, in which IFN is widely used for treating various diseases successfully, benefiting many patients and also procuring handsome profits for the manufacturing companies. Biology and medicine are now making remarkable progress by the efforts of many people including the speakers and other participants of this symposium, and we can certainly foresee a bright future in many areas. But I leave the task of describing such a future to other people. What I want to present here is a dawn to something still uncertain, poorly defined.

In spite of the great progress expected, or even promised, for the currently ongoing biological and medical research, I don't feel very

comfortable about the direction to which our mainstream biology and
medicine are leading us. Why not? Biology acquired a respectable
position as natural science comparable to physical science only relatively recently. The advent of molecular biology in the 1950s contributed greatly to this. The progress of biology was due largely to the
reductionist methodology, seeking explanations of living systems by
reducing them into molecules and other elementary units. This has been
a highly successful methodology, but it inevitably leads us to a mechanistic view of living things, the view that living things can be totally
explained in terms of matter and the physical causality relationships; in
short, living things are understood as machines. This trend was expressed by Webster and Goodwin [26], the proponents of the so-called
structuralist biology, as disappearance of living organisms. They wrote:
"Organisms have disappeared as fundamental entities from modern
biology, replaced by genes and their products as the primary determi-
nants of selected characters."

Many biologists naturally resisted this trend, and supported a view of
life which is generally called organicism [16]. Organicism is against
reductionism, but in the final analysis, it is a kind of mechanicism, as I
discussed elsewhere [14].

That biology is mechanistic has its root in the very premise of
modern science that cuts object from subject, body from mind, and
leaves out subject and mind from considerations. So, long before organisms disappeared from biology, subjectivity of living organisms was
eliminated from its domain. This was perhaps necessary for our biology
to become a full-fledged natural science, but it leads to distorted images
of living things.

How about our medicine? It has made great progress, especially since
the early twentieth century, by adopting scientific methods, by seeking
objectively observable facts and the physical causality relationships
among them. This was certainly welcome, because the remedy provided
by modern medicine was in many cases reliable and widely applicable
to patients regardless of their social and cultural backgrounds, but it is
also true that modern medicine causes various conflicts with patients'
interest. Simply put, what patients need is care, whereas what doctors
provide is cure, and they often don't agree with each other. Such conflicts seem to me inevitable as long as our biology, on which our
medicine is based, is as mechanistic as it is today.

I don't deny the value of contemporary biology and medicine at all. On the contrary, I look forward with excitement and respect to seeing their development, but I strongly object to the idea that they are the only reliable and correct ways of knowing about living beings and dealing with human health. I believe we need a biology dealing with really living organisms, and for that, we must admit their subjectivity. I hope semiotics will provide a scientific way of doing that. Semiotics is the science of signs, which comprise many varieties, such as symbol, signal, symptom, and so on. It is concerned in large part with human culture, including areas like language, art and literature, but it will be easy to see sign processes widely occurring also in the non-human living world. In the following, I will first describe how I was led to the biosemiotic view of interferon.

5.2 Analogy Between the Cytokine System and Language

My wanderings in thought began in the late 1980s when I became aware of a close analogy between the cytokine system and human language. There are many species of cytokines, and they interact with each other to form a network as a whole. Each of them has multiple biological activities, but can exert a particular action depending on the kind and the condition of the cell it acts on. In this respect, a cytokine molecule looks similar to a word in language, since a word usually has multiple meanings, and can indicate a particular meaning depending on the context. Like cytokines, there are many words in a language that are related to each other to form a consistent system as a whole. Various other aspects of the cytokine network can be noted to be similar or isomorphic to language. I summarized this thought in a lecture I gave in Florence in 1989 as a newly elected honorary member of the International Society for Interferon Research [10]. I pointed out there that as a word is a sign that represents something in the world (or in the brain), cytokine molecules are the signs that represent the condition to which the organism is subjected. Namely, certain events or conditions inside and outside the body will induce certain cells to produce pertinent cytokine molecules, which serve for the target cell as signs of those events or conditions. Then in the target cell, the signs are decoded by what we could call the homeostatic code to produce relevant physiological effects.

Analogy to language has been noted in a different area of biology for many years, that is, in molecular genetics. Thus, a genetic message is written linearly with the four-letter nucleotide alphabet, and is translated into the protein language consisting of a linear sequence of the 20-letter amino acid alphabet. The nucleotide sequence in DNA is hierarchically organized at several levels, like a verbal message constructed by a linear organization of words and other linguistic units at a number of hierarchical levels. This remarkable similarity to language is, however, essentially of the nature of syntactics; that is, the question of what are the units of the system at different levels and how these units are arranged to yield meaningful structures. The analogy to language noted above for cytokines is of an entirely different nature and belongs to semantics, that is, the realm of meaning. We are concerned here with a physiological language of our body, with cytokines as its semantic units.

Our body has another important system of intercellular communication, that is, the nervous system, which utilizes signalling systems diametrically different in nature from those of cytokines and hormones. With these two systems working together, each cell in our body can have representations of the worlds inside and outside our body.

Such a view was, in retrospect, nothing special, because, from the outset, everybody recognized cytokines as vehicles for cell-to-cell communication in our body. But it is important to note that we use the word "communication" to mean, not just transmission of physical stimuli, but transmission of messages that contain certain meanings for the cell. Namely, my main point was to regard the cytokine molecules not simply as physical entities, but as entities that carry meanings.

Now, words in a language are a particular kind of sign. A 'sign' is defined generally as something that represents or stands for something else. So I thought, cytokine molecules representing certain physiological conditions are, like words representing some meanings, qualified as a kind of signs.

However, an important characteristic of signs in general is that the linkage of a sign to its meaning is arbitrary and is determined by convention, or more specifically, by code. For instance, a word written or pronounced as dog or Hund or chien or inu is not linked with necessity to its meaning, that is, a kind of four-legged animal. Therefore, if the world of molecules is governed strictly by physicochemical principles, allowing no room for arbitrariness, then cytokine molecules may not be called signs. However, we know well that the action of a cytokine can be altered artificially, if the receptor of the target cell is appropriately manipulated by genetic engineering. That is, the biological activity of a cytokine is not an intrinsic character of the molecule derivable directly from its physicochemical properties, but is determined by the kind of cell it acts on. So the linkage between the cytokine molecule and its activity must be considered to be arbitrary, as it is determined mainly by biological principles containing historical contingencies. The situation is very similar to language, in which the linkage of a word to its meaning is arbitrary and determined mostly by historical contingencies, although, for the members of the society, the word-meaning linkages may appear to be unchangeable and far from arbitrary. So I concluded cytokine molecules could be called signs.

Then, after several years of musing, I came to realize that not only cytokines, but protein molecules in general, and further, any molecules of high and low molecular weight, and any ions, that constitute a living system can be understood as signs. I was thus led to a general biosemiotic view of living beings [11,12].

5.3 Biosemiotics

As I said before, semiotics is the science of signs, and its major concerns are signification and communication, that is, how signs are produced and how they are communicated within a society. As everybody knows, animals and birds make good use of various kinds of signs in their life, that is, they make noises, sing songs, make gestures, display wings etc., when they seek mates, search food, and so on. So, animal ethology heavily (if not entirely) overlaps biosemiotics, as the pioneer of biosemiotics Sebeok [18,19] pointed out many years ago.

Semiosis, that is, sign process, occurs not only in macroscopic systems, but also in microscopic and molecular systems. Thus, a pheromone molecule is a sign that represents a mate for its partner; a cytokine or a hormone molecule represents the condition or the event which the organism is experiencing.

Semiosis usually involves the sender and the receiver of the sign, and the important point is that the concept of sign presupposes the presence at least of the receiver with the capacity to interpret the sign, to understand its meaning. Therefore, the receiver must be a living being (or its part or extension). The sender of the sign, on the other hand, may be an inanimate body. Thus, a dark cloud growing rapidly serves for a farmer and other observers as a sign of rain coming soon. The receiver of a sign confers a meaning on the sign, according to a certain code or convention, but no mental faculty is necessary for the participants of sign processes. An interesting idea of molecular signs was proposed by Tomkins [24], a biochemist famous for his studies on the mechanisms of hormone actions. When *E. coli* cells are starved of a carbon source, the cell accumulates cyclic AMP, which then down-regulates various metabolic reactions. Here, the cyclic AMP molecule is a symbol (a kind of sign) representing shortage of carbon source in the environment, and the symbol is interpreted by the intracellular machinery of the bacterial cell according to the metabolic code so as to adapt the cell better to the poor environment.

Encouraged by this paper and also by more recent papers by Hoffmeyer and Emmeche [6] and Emmeche [4], I came to realize that the idea of molecular semiosis is applicable not only to molecules specialized for communication, such as cytokines, pheromones and neural transmitters, but to molecules in general that constitute living systems [11,12]. Take an enzyme molecule, for example. Its reaction with its substrate is a physical process, but when it is assembled, together with other molecules, into a metabolic circuit in a cell, the enzyme molecule becomes an integral part of the system that produces a useful substance for the cell, and thereby it acquires a role or a meaning that is specified by the system it belongs to. If the same enzyme is integrated in a different metabolic circuit, it acquires a different meaning as specified by the new system. The same is true with cytokines, as I mentioned before. Namely, a cytokine molecule has a certain meaning that does not arise from the molecule itself but is conferred upon it by the system it belongs to. This can generally be said for any molecule that constitutes living systems.

One tends to think that the world of molecules is governed by the principles of physics and chemistry and allows no room for semiosis. But actually, even at the molecular level, there are various kinds and degrees of freedom and indeterminacy as regards the chemical reactivity of a molecule, and also as regards the choice of the molecular composition to construct a system for a given purpose [12,21]. I won't go into detail here, but there are reasons to believe that living systems are composed of a set of molecules and a set of relations among them, selected, without strong chemical necessity, out of numerous physically possible combinations, according not solely to physical causality relationships, but to semiotic principles appropriate for the purpose of the system.

So living things can be regarded to consist of assemblies of sign processes at all levels of hierarchy, from molecule to cell, to organism, and to ecosystem. The interactions among the molecules, the cells and the organisms that eventually form the biosphere are not only physicochemical, but more importantly, semiotic in nature. Therefore, the term semiosphere will be more appropriate than the term biosphere to represent the total living world [5].

In the biosemiotic view, there is semiosis wherever there is life. The living state is realized based on semiosis, and semiosis is the very feature that distinguishes life from non-life. Then a living being is an entity that maintains its integrity by interpreting the world, creating meanings out of it, and assembling the elements in a teleonomic manner. In this view, every living thing, even a unicellular organism, must be considered to have its own autonomous subjectivity, regardless of the presence or absence of mental faculties. I will come to the topic of subjectivity again later.

In our orthodox science, a molecule is a molecule and should be fully described in terms of physics and chemistry, and it will therefore be against the Occam's principle of parsimony to confer upon it such a superfluous character as being a sign. My answer to this is that I use the word sign, because it is an entity with meaning, and because I consider the most important thing about a molecule is the meaning it has for the living system it belongs to. Calling a molecule a sign has a merit of implicating that a certain higher system is inherent in that molecule.

One may rightly point out that my biosemiotic view is anthropomorphic. Anthropomorphism is generally believed to be something that has to be avoided strictly in our science, but it was molecular biology around the middle of the twentieth century that began to introduce numerous anthropomorphic terms into biology and the world of molecules, starting from the all-important term "genetic information", followed by recognition, regulation, messenger, translation, and so on. These terms are usually taken as metaphors, which one can replace, if one wants to, by physical terms allowed in natural science. But I don't think anybody succeeded, or will succeed in future, to replace them with physical terms without losing their biological meanings. I believe these terms represent certain essential characters of biological molecules, and are therefore indispensable for precisely describing their behavior. They are reflections of the purposiveness and semiotic character of the molecules in living systems. Molecular biology is often considered to have succeeded in reducing biology to physical science, but the presence of so many anthropomorphic terms indicates that actually the presumed reduction is impossible. Reduction was certainly carried out, but it was not into purely physical molecules but to semiotic molecules. So, molecular biology appears in a large measure to be biosemiotic.

Here, I would like to divert a little, and direct your attention to some general problems that our biology and medicine have to face.

5.4 Biology at the Whole Organism Level

If we look back on the development of biology during the latter half of the twentieth century, we will first note that the advent of molecular biology in the 1950s brought about profound changes in the whole field of biology. "Molecular thinking" penetrated every field of biology, and many investigators became to prefer dealing with molecules to dealing with biological or clinical phenomena, because molecules were easier to handle and gave better reproducibility. Then, in the 1970s, there occurred a subtle shift in the focus of research interest from the molecule to the cell, and the term "cell biology" came to be frequently used in place of molecular biology. A symbolic event was the initiation of that prestigious journal *Cell* in 1974; also, the great textbook *Molecular Biology of the Cell* by James Watson and others appeared in 1983 [1]. The star players in the age of molecular biology were *E. coli* and its phages, and those in the age of cell biology were eukaryotic cells.

Much of current research in cell biology, concerned with such problems as cancer, immunity, homeostasis and embryogenesis, aims actually at understanding the events that occur in whole organisms of higher living forms. So, it seems that our research interest has already shifted from the cell to the whole organism. In other words, we seem to have returned to classical biology, in that our major concern lies in what happens in the macroscopic living organisms, although actual analyses are still largely directed to the cell and molecular levels. In various areas, a gap is felt between the demand of the time and what we can practically do with the available methodologies. For instance, we now have the complete base sequences of the whole genomes of various species, but how this wealth of information can be put to good use to understand the living organisms remains largely left to be investigated.

This applies also to IFN research. IFN and cytokines are the agents that work meaningfully in the context of whole organism. We now know very much about the properties and actions of many cytokines at the cell and molecular levels, but we don't know so much about the cytokine network and its workings in vivo. Clinical uses of IFN are certainly bearing fruit, but much more should be gained if we knew better about how IFN works in vivo. Perhaps we still need many more empirical observations at the whole body level in the clinics as well as in the laboratory. Some IFN investigators are reporting on an intriguing possibility that oral administration of small amounts of IFN exerts significant effects on human and animal bodies. The effects, if real, may well be influenced by a great many factors, and so we may have to develop new methodologies effective for such investigations of highly variable multifactor systems.

Needless to say, problems at the whole organism level can be approached without biosemiotics, but generally speaking, sign processes become more important as the level of organization goes higher in hierarchy, because the degree of physical determination will generally go down.

Here, I would like to emphasize that, in order to understand what happens in the whole organism, one cannot avoid asking how the mind is involved in the phenomena one studies. It is now well recognized that the cytokine system is intimately interwoven with the nervous system. How the functioning of the IFN and cytokine system is linked to our mental or psychic factors is a topic of great clinical and theoretical interest. It is noteworthy that our biology often seeks to reduce mental factors into physical substances and parameters. For instance, investigators will look for possible correlations between certain mental states or emotions and such physically observable changes as the level of certain effector molecules or the NK cell activity, and try to explain the effects of mental activities by such physical parameters. These investigations are undoubtedly valuable, but mental activities are after all not totally reducible to matter and physical parameters, because they are sign processes, although of course they are based on material processes. Our brain is not only a physical system but, more importantly, a semiotic organ that creates and manipulates all sorts of signs; our thoughts and feelings are semiotic events, and they need be approached as such. So there is the question of how we can incorporate mind into scientific inquiries without totally reducing it into matter. I regret I can say nothing at present, except perhaps that we will need extensive systematic clinical and laboratory observations at the whole body level about the effects of mind on cytokine and other bodily functions.

5.5 Disease vs. Illness

My next topic is sickness, and the attitude of our medicine to it.

A sickness is detected by symptoms and signs. It has both subjective and objective elements; to distinguish them, sometimes a "symptom" is defined as the subjective evidence perceived and described by the patient, while the objective evidence perceptible to the doctor is designated to be a "sign". At any rate, sickness is clearly a topic that needs be studied, not only in medicine, but in semiotics too. Also, I would like to point out that although sicknesses usually look like personal events or processes that take place naturally in an individual, they are defined and dealt with under strong influences of social and cultural factors, and so they are in a large measure social institutions. In our societies, we tend to confine sickness into the realm of medicine, dissociated as a profession from semiotics, sociology and other realms. As a result, the social nature of sicknesses does not seem to be receiving necessary attention.

Some medical anthropologists separate sickness into two categories: disease and illness [22,23]. Disease is the occurrence of objectively verifiable changes within the body, i.e., the physiological-biochemical processes, while illness is the social and personal experience of bodily or emotional dysfunction. So the patient suffers from an illness, whereas the doctor treats a disease; illness is something to be cared and healed, whereas disease is to be cured. The target of modern medicine must be the disease, as long as it aspires to be science-based, whereas the patient wants to have its illness cared and healed. So the conflict is not rare between the patients' interest and the medical measures they receive. I know the discordance between care and cure is being increasingly recognized, and various measures are being developed. An easily visible case is that of dying cancer patients, for whom humane care of personal and social nature is often given priority over science-based medical measures against their disease. This is not a special case. For various sicknesses, it is often hard to decide how to strike a good balance between care and cure. For example, anxiety is a very common mental state, which can often develop into an illness or a disease. Modern medicine will be able to cure the disease by using psychotropic drugs, which will also heal the illness. This is welcome indeed, but, if we eliminated all anxieties and other mental hardships by means of drugs, what kind of a world would it be? I certainly don't want to live there.

I know we still need to clarify the material bases of various mental and bodily diseases, but I also know that sicknesses are not totally reducible to matter, to physiological-biochemical processes. So, sicknesses must be studied, not only in medicine, but, more importantly, also in sociology and the humanities. But the temptation to reduce medicine into material science is in general very strong.

Biology seems to be generally believed to be the foundation of all medicine. I don't doubt that biology-based medicine is of great value, but if one thinks of the problems of AIDS, it will be immediately evident that social and human sciences are at least equally important to deal with the illness as compared with medicine which tries to cure the disease. An interesting and encouraging incident was the total eradication of smallpox from the earth, achieved about 20 years ago by the worldwide campaign led by WHO. The smallpox vaccine was introduced to the West 2 centuries ago by Jenner, and used in the campaign with some technical improvements. The power of medicine was certainly indispensable there, but the decisive factors for the success were the social, political and administrative wisdom and actions. This case thus testifies that even relatively simple medical techniques can achieve great things if combined with appropriate social, political and economic measures.

5.6 Conceptual Structure of a Living Organism in Biosemiotics

With the discussion of the above two topics, I hope I could convey to some extent my feelings about the inadequacy of our present-day biology and medicine. Broadly speaking, the inadequacy seems to be deeply rooted in the attitude of modern science toward the world, which is reductionistic-mechanistic-atomistic. Such an attitude may be readily accepted by the individualistic tendency of the Western mind, and is reflected in the basic image of a living organism in our mainstream biology and medicine. Namely, the image seems to be that of an independent, solitary individual, which can in principle sustain its autonomous living by making appropriate use of its environment. A typical example is the image of a bacterial cell in molecular biology. There, a single cell can grow and multiply in a medium, utilizing the nutrients in the environment, and no interaction with other cells need be considered. With higher living forms, an individual organism will interact with others when needs arise, such as mate-seeking and food-hunting, but until such needs arise, it will make its living freely as an independent individual. Humans will form societies because of their high intellect, but other living things will do so only to the extent that is necessary for making their living. Even when a non-human species forms a society, it will exist only in a primitive form.

I think such a view of independent individuals is fundamentally incorrect. Against this fallacy of the atomistic, self-sufficient individual of living beings that appears to prevail in our biology and medicine, I propose a quite different image of a living organism based on biosemiotics.

To present that image to the reader, I have to discuss first about our notion of the environment of living things. We tend to think that the physical world we perceive is the only real world, and all living things live in this same physical environment. However, as Jacob von Uexküll, a German animal physiologist, advocated in his famous *Bedeutungslehre* [25] and other books more than half a century ago, each living being has its specific world, defined by its cognitive abilities and action faculties. He used the word *Umwelt* to designate this species-specific world. For instance, a tick living in the woods makes its living by climbing up a tree, sensing the odor of butyric acid from a warm-blooded animal coming nearby, falling on it, feeling its warmth, sucking the blood, laying its eggs, and so on. This relatively simple set of the objects of perceptions and actions constitutes the whole world for the tick, and that is its *Umwelt*. It is a subjective world specific for the organism. It is a semiotic concept, as it is the world of meaning.

In the mainstream biology, it is usually stated that a living being is adapted to a certain niche, but this is somewhat misleading in that it gives an impression that the organism and the physical environment are independent of each other and the latter has a one-sided determining power over the former. Actually, however, the living organism and the environment, or rather, the *Umwelt*, are closely intertwined, and they coexist and evolve together. It is true that, ever since life emerged on the earth, living forms gradually acquired relative independence from the physical environment through evolution, but they will never stop depending and acting on the environment and forming their *Umwelts*. So we must regard the *Umwelt* as an inherent element of a living organism.

Another aspect of the biosemiotic image of a living organism is the sociality inherent in it. Living organisms cannot live a solitary life, but must live by interacting in various ways with other organisms, helping each other and competing for resources. This is true regardless of whether the organism is a sexual or asexual one, and regardless of whether it lives individually in a dispersed state or in a group with other individuals. Even bacterial cells, which may appear to be typical solitary organisms, have been shown to interact with each other extensively, exchanging genetic information and differentiating into various cell types [15, 17,20]. Also, many cases of their interaction with organisms of other species are known, resulting in symbiosis. These interactions are realized by reciprocal molecular conversations. In various bacteria, the so-called secondary metabolites of numerous kinds are produced which are not essential for their growth. It seems likely that these chemicals serve for conversation with other organisms of the same and different species [15]. Thus, forming a society is a common denominator of all living things from bacteria to human beings.

It may be reasoned a priori that an individual that interacts with others and changes its state accordingly in a flexible manner will have a greater chance of survival and evolutionary diversification than an individual with less flexibility.

When an organism is born, it is not born into an arbitrary physical environment, but into its specific *Umwelt*, suitable for its living, and it starts its own life in a style it inherits either innately or by learning from the progenitor. So both the *Umwelt* and the life style is the legacy of its progenitor. When I say progenitor, it does not mean a simple sum of the individuals along the ancestral line, but the totality of interactions between the individuals and between the individual and the *Umwelt*, that is, what is to be called the species society in its historical context. Thus, an individual lives from the outset in its species society.

But what is a society?

Various species of ants and bees are called social insects, because they live in groups of many individuals, which are differentiated into queens and workers with various different tasks. It seems widely accepted to call their assembly a society. Even assemblies of certain single cells are sometimes called cellular societies, perhaps metaphorically, if they interact with each other and mutually modify their characters in meaningful ways. On the other hand, if, for instance, you arrange many robots in good order in an assembly line, they may cooperate with each other and produce useful products, but you will never call the assembly a society. Why not? It may simply be because the elements are inanimate things and are not alive. This will appear a good reasoning, but I don't think being alive or not is a satisfactory criterion for the elements to form a society. Instead, the necessary criterion is, I believe, that the elements have their own autonomous subjectivity. Thus, we form societies because each of us is a subject and not a passive object. Insects form their society because each member is in itself a subject, however limited its subjectivity may be compared with self-sufficient organisms. If every living organism has its inherent sociality as I asserted above, then every organism must have its autonomous subjectivity.

To be alive and to be a subjective entity are not equal. Society is normally defined to consist of individuals of the same species, but the individuals of a given species usually live together with organisms of many different species. Some of them may stand in symbiotic relations. Then, they are the necessary members of the society, and therefore they are subjective entities. But other living organisms which are, for instance, the prey or predator of the species are not subjects but objects, and are not members of the society. If you kill a cow and eat it, the cow is an object, but if you keep a cow and love it and the cow loves you, it becomes a subject and stands in a social relation to you. So the subjectivity of an organism appears to be the satsifactory criterion of its sociality.

Summarizing these considerations, a living organism is understood to exist as a triadic structure, consisting of, first, the individual, that is, the subject, second, the *Umwelt*, and third, the society [14]. These three elements are linked inseparably to each other, and the linkages are mediated by semiosis. The picture of the atomistic individual of a living organism standing alone in a physical world is possible only as an abstraction, justifed by the Cartesian cut of subject and object in modern science.

The concept of subjectivity has not been allowed to enter into the domain of natural science, because natural science should be concerned only with objectively observable facts. Also, subjectivity is generally associated with mental faculties, and Western thinking traditionally tends to allow only humans to possess mind.

The so-called classical ethology by such prominent investigators as Konrad-Lorenz and Tinbergen tried to avoid referring to mind of animals, and instead took recourse to stimulus-response and other automatic mechanisms to explain animal behaviors [2]. This is a very artificial attitude, similar to the behaviorist psychology, reducing animals to objects rather than regarding them as subjects. Recent studies of animal behavior by, for instance, Japanese primatologists [8,13] and Frans de Waal [3] began to clearly indicate inadequacy of such attitudes.

Criticism of a similar nature can be directed to sociobiology. The concept of society of non-human living beings was slow to be generally accepted as a legitimate agenda in biology. So, the advent of sociobiology in the 1970s was a notable event, but the method of sociobiology of E. O. Wilson [27] and others was clearly reductionistic, seeking explanations of animal sociality in terms of genes and in terms of physical causality relationships among the elements. Again, animals here are objects and not subjects, and the so-called society looks like an intricate mechanism that works automatically. This discipline is certainly useful in elucidating certain aspects of living things, but to understand their societies, we clearly need sociological approaches, admitting subjectivity of the constituent individuals. What we really need is biosociology, rather than sociobiology, as propounded many years ago by Kinji Imanishi [7], a Japanese biologist and guru, who, among other things, initiated studies of primate societies along the sociological lines of thought [13].

We can admit the presence of mind for some animals, but obviously not for all living things. Instead of mind, or psyche, or soul, or spirit, we can point out subjectivity as a common denominator of all living things from *E. coli* to humans. A subject is an agent that acts on the *Umwelt* and the society through semiosis, and thus makes its living. It is an entity that interprets the world, creates meaning out of it, and responds selectively to it. It is a self-referential agent which incorporates the past into the present and the present into the future [5].

Is my view too anthropomorphic? I admit it is anthropomorphic, and I believe we need to be so in describing certain essential aspects of living organisms. Modern biology tried to avoid anthropomorphism, but it was not totally successful in that molecular biology needed so many anthropomorphic terms and concepts. Besides, our biology was seized by mechanomorphism in its attempt at avoiding anthropomorphism, seeking mechanical explanations based on matter and physical causality.

If I ask you, "Are we machines?", you will probably answer "No" with confidence. But what is the reason for the negative answer? If you confine your reasoning to the realm of natural science, I am not sure if you can give a satisfactory answer. Now that the molecular mechanisms of cellular processes have been clarified in considerable detail, the view that a cell is a machine does not seem extraordinary any more. Multicellular organisms are certainly highly complicated, but aren't their complexities a matter of degree after all? I myself cannot propose theoretical reasons why the morphology and function of any living form cannot ultimately be explained in terms of matter and physical causalities.

Then, to the question "Are we machines?", the honest answer from our orthodox biology will have to be "yes". In other words, our biology is of such a character that it extracts the features of living things as machines and studies their mechanisms. Theoretically, therefore, the only picture of living things we can expect from our orthodox biology is that of machines. I am not denying the value of our contemporary biology and its future extensions, I am not against its mechanism, but I am against the view that our mainstream biology, and for that matter, natural science in general, is the only reliable way of getting knowledge about our world.

I believe we are not machines simply because we are autonomous subjects. Machines are not alive, not because they are made of inanimate materials, but because they are not subjects in themselves. Modern science has avoided to deal with subjectivity, but such an attitude may not be sustainable any more in various areas. For instance, in brain science, subjective experiences like the sense of pain will have to be studied, and such philosophical problems as mind-body dichotomy will become a legitimate agenda. Recent developments in the so-called complex systems and artificial life are pointing to a similar direction. In a recent book [9] describing mathematical models and computer simulations of evolution of life based on complex systems science, the authors write that one cannot leave out "meaning" without reducing the theory to mere mathematics; in constructing artificial systems to simulate life, the constituent elements must have internal degrees of freedom, and they are most aptly called "subjects". The fact that these mathematicians-physicists find such concepts as meaning and subjectivity indispensable for their work reflects a growing recognition in various areas that the methodology of modern science is not effective. So, it seems we are in a dawn to a new kind of science, in which, I hope, really living organisms will have their place to live in happily.

References

1. Alberts B, Bray D, Lewis J, Raff M, Roberts K, Watson JD (1983) Molecular biology of the cell. Garland, New York
2. Crist E (1998) The ethological constitution of animals as natural objects: the technical writings of Konrad Lorenz and Nokolaas Tinbergen. Biol Phil 13:61–102
3. De Waal F (1996) Good natured. The origins of right and wrong in humans and other animals. Harvard University Press, Cambridge, MA
4. Emmeche C (1991) A semiotical reflection on biology, living signs and artificial life. Biol Phil 6:325–340
5. Hoffmeyer J (1996) Signs of meaning in the universe, trans. by BJ Haveland. Indiana University Press, Bloomington (Advances in semiotics)
6. Hoffmeyer J, Emmeche C (1991) Code-duality and the semiotics of nature. In: Anderson M, Merrel F (eds) On semiotic modeling. Approaches to semiotics, vol 97. Mouton, Berlin, pp 117–166

7. Imanishi K (1941) Seibutsu no sekai (The world of living things). Reprinted as Koudansha Bunko, 1972. Koudansha, Tokyo
8. Itani J (1985) The evolution of primate social structures. Man 20:593–611
9. Kaneko K, Ikegami T (1998) Fukuzatsu-kei no shinka-teki scenario (Evolutionary scenario of complex systems). Asakura, Tokyo
10. Kawade Y (1990) Cytokine network in analogy to language – a general view of interferon research from a distance. J IFN Res 10:101–107
11. Kawade Y (1992) A molecular semiotic view of biology. Interferon and 'homeokine' as symbols. Riv Biol (Biol Forum) 85:71–78
12. Kawade Y (1996) Molecular biosemiotics: molecules carry out semiosis in living systems. Semiotica 111:195–215
13. Kawade Y (1998) Imanishi Kinji's biosociology as a forerunner of the semiosphere concept. Semiotica 120:273–297
14. Kawade Y (1999) The two foci of biology: matter and sign. Semiotica (in Press)
15. Losick R, Kaiser D (1997) Why and how bacteria communicate. Sci Am 276(Feb): 52–57
16. Mayr E (1997) This is biology. The science of the living world. Harvard University Press, Cambridge
17. Miller VM (1998) Bacterial gene swapping in nature. Sci Am 278(Jan): 47–51
18. Sebeok TA (1972) Perspectives in zoosemiotics. Mouton, The Hague
19. Sebeok TA (1989) The sign and its masters. University Press Am, Lanham
20. Shapiro JA (1988) Bacteria as multicellular organisms. Sci Am 258(Jun): 62–69
21. Sibatani A (1990) Stability of arbitrary structures and its implications for heredity and evolution. Wissenschaftkolleg zu Berlin Jahrbuch 1988/1989, pp 206–217
22. Staiano KV (1979) A semiotic definition of illness. Semiotica 28:107–125
23. Staiano KV (1992) Biosemiotics, ethnographically speaking. In: Sebeok TA, Umiker-Sebeok J (eds) Biosemiotics. The semiotic web 1991. De Gruyter, Berlin, pp 407–426
24. Tomkins G (1975) The metabolic code. Science 189:760–763
25. Uexküll J Von (1940) Bedeutungslehre. English edn (1982) The theory of meaning. Semiotica 42:25–82
26. Webster G, Goodwin B (1996) Form and transformation. Generative and relational principles in biology. Cambridge University Press, Cambridge, England
27. Wilson EO (1975) Sociobiology: the new synthesis. Harvard University Press, Cambridge

6 The Clinical and Social Impact of Interferon-β: The First Approved Therapy in Multiple Sclerosis

H.F. McFarland

6.1 Introduction

The approval of interferon-β 1b for the treatment of multiple sclerosis (MS) in 1993 has had a profound impact on the attitude of both physicians and patients concerning the illness. To fully appreciate the impact one needs to understand the failures of previous therapies and the frustration associated with dealing with a disease of uncertain prognosis and course. The approval of interferon-β 1b opened a new era of treatment of and research on MS. In this review I will examine the natural history of the disease in order for the reader to understand the difficulties associated with the care of patients with MS and in identifying new therapies for the disease. I will then discuss the impact of that approval and the current thoughts on treatment MS.

6.2 Natural History of MS

Clinically, MS is characterized by the absence of a predictable course. However, some generalizations regarding the clinical course have been drawn from studies of natural history cohorts [1]. In most cases the disease begins with a course that is characterized by acute episodes of neurological dysfunction followed, even without treatment, by improvement. These episodes are known as relapses or exacerbations and the course is termed the relapsing remitting course of MS. In a majority of cases, over time, the relapses will often be followed by less than complete recovery with accumulation of disability following each attack. Eventually, in many cases, progression of neurological dysfunction will occur independently of the acute episodes of worsening. This phase of the disease is known as the secondary progressive course. A small number of patients will have a progressive disease from the onset and if exacerbations are absent the course is termed primary progressive MS. Overall about 90% of patients begin with a relapsing remitting course. Of these about two-thirds will go on to a secondary progressive course and about one-third will continue to have a relapsing remitting course. One of the frustrations of MS is the inability to predict at the onset which patients will have a progressive course and, of those, which will accumulate sufficient disability over time to require them to substantially modify their life. The most important measure of the disease is the level of disability that occurs of long term follow-up. Again, based on natural history studies followed for 15-20 years, about one-third of patients will have only mild involvement which will not lead to significant impairment of their activities of daily living. That means that while they may have some symptoms, they will be able to perform nearly everything that they wish to do. Another third will have a more significant disability and will have some limitation of their activities but will still be able to perform most of their daily activities with only minimal assistance. The last third will have more marked disability and will have limitations on their ability to walk and will generally be limited to a wheelchair or possible bedridden. Only a small number of patients have a truly aggressive course. MS generally does not shorten life but in a very small percentage of individuals MS will lead to death over after a short, aggressive course of a year or less. Until just over 10 years ago when magnetic resonance imaging (MRI) started to be used to study

MS, it was generally thought that in the early relapsing remitting phase of disease that the disease was inactive between exacerbations. Studies using MRI has shown that this understanding of the disease is wrong.

6.3 Implications of MRI on Understanding the Natural History of MS

Most studies of MS using MRI have used two conventional imaging techniques; T2-weighted images and T1-weighted images done after the administration of gadolinium DTPA. The importance and value of these two imaging techniques is best examined from the standpoint of what is thought about the pathogenesis of the MS lesion [2].

In general the MS lesion can been considered as occurring in three different steps. The initial step of the MS lesion is characterized by inflammation. Subsequently, immunological mechanisms amplify the acute inflammatory response and mechanisms responsible for damage to the myelin sheath are initiated. The acute MS lesion is rich in macrophages containing myelin debris and an active edge of myelin destruction can be seen. Eventually the acute inflammatory aspect of the lesion resolves and in its place are circumscribed areas of myelin loss with variable degrees of axonal damage and gliosis and some residual inflammatory cells. Each of the stages of the MS lesion is associated with an increase in tissue water. The acute lesion is edematous while the chronic lesion has either increased extracellular space secondary to loss of myelin and/or axons or increased intracellular water associated with gliosis. On T2-weighted MR images increased concentration of protons or water results in increased signal. Thus, the MS lesion, regardless of age will be seen on MRI as an area of increased signal. Consequently, T2-weighted images, while pathologically non-specific, are sensitive for demonstrating the changes in the nervous system associated with MS. Because of the sensitivity of T2-weighted images to detect areas of disease in MS, the technique has provided a very valuable tool for assisting in the diagnosis of MS. Early in the use of this imaging technique to study MS it was noted that new lesions occurred in patients who were clinically stable.

The extent of disease activity in patients during the early relapsing remitting phase of the disease and who are not having exacerbations has been more extensively studied using contrast enhanced images. Contrast-enhanced images are obtained using a T1-weighted sequence done following the administration of gadolinium-DTPA (Gd-DTPA). Gd-DTPA is a paramagnetic contrast agent that causes shortening of the T1-relation times which produces an increased signal on the T1-weighted image. Gd-DTPA does not cross the intact blood brain barrier (BBB). It is thought that the initial step in the developing MS lesion is migration of an activated T-cell across the BBB. If this T-cell encounters an antigen to which it is sensitized, the T-cell begins a cascade of proinflammatory cytokines that both initiates and then amplifies the inflammatory response; one component of this process is disruption of the BBB allowing an influx of inflammatory cells. Thus, lesions identified following the administration of Gd-DTPA will represent lesions in their early, acute stage. Several groups of investigators have now applied contrast-enhanced imaging to study the natural history of disease activity during the early relapsing remitting phase of MS [3–5]. In a cohort of 154 patients who were imaged for a minimum of 3 months serially, over half had at least one enhancing lesion on the initial MRI and nearly 70% had enhancement on one of the 3-monthly images. Thus, the level of activity as measured by the occurrence of new, active lesions is high even though this cohort of patients had mild disease and was studied during periods in which they were clinically stable. Smaller group of patients with the same characteristics have been followed monthly by serial enhanced MRI in some cases for at least 12 months and found to have an ongoing activity [4, 6,7]. Although the number of new lesions fluctuates from month to month, in patients followed serially for 12 months or longer, the average frequency of contrast enhancing lesions determined over several months will be relatively stable within individual patients (Fig. 1).

Overall these findings indicated that MS is an active process in most patients from the onset and this additional understanding of the level of activity occurring during the early phase of the disease while patients are clinically stable has added to the frustration of dealing with the disease. While the relationship between disease activity on MRI and that seen clinically has been imperfect [8], the relationship between the extent of disease seen on MRI at the time of initial presentation with signs or symptoms suggesting MS (clinically isolated syndromes) and the likelihood of developing clinically definite MS 5 or 10 years later

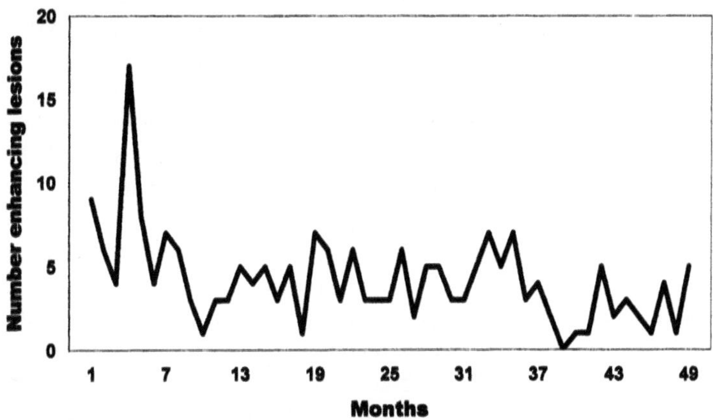

Fig. 1. Number of contrast-enhancing lesions per month. Patient with relapsing–remitting multiple sclerosis and mild disability

has been high [9,10]. Nearly 90% of the patients presenting with a clinically isolated syndrome and with several areas of increased signal on a T2-weighted MRI developed clinically definite MS over 5 or 10 years compared to approximately 10% of patients with normal MRIs at presentation. In addition, the extent of disability occurring over 5 or 10 years correlated with the degree of MRI abnormality seen at presentation. A similar relationship has been found in follow-up studies of patients presenting just with optic neuritis [11]. These studies did not explore the predictive value of contrast enhanced MRI but since almost all enhancing lesions result in a residual T2 abnormality, it would be expected that patients early in the disease course with contrast enhancing lesions will tend to have a course associated with greater disability than patients with inactive MRIs. Consistent with this hypothesis is the observation that exacerbations are more frequent in patients with high numbers of contrast enhancing lesions [4,12].

What do these observations mean to the physicians caring for patients with MS and have the implications changed after the approval of IFN-β 1b? The attitude of physicians seeing patients with early MS can be viewed in three phases. Prior to the extensive studies of MS using MRI, physicians seeing a patient early in the disease had little information that would allow the doctor to distinguish patients likely to accumulate significant disability with time from those expected to have a more benign prognosis. Thus, the physician could be optimistic that the patient would be one with a mild course and the real frustration in dealing with the disease would not occur until the patient really began to accumulate disability. As the data obtained for MRI studies of RRMS began to emerge the frustration began to occur earlier in the care of patients. Physicians seeing a patient early in the disease and without disability but with significant disease on MRI and especially with active contrast enhancing lesion knew that the possibility that the patient might not do well over time was high. However, the physician had no reasonable treatment options that might modify this course. In addition, testing of new therapies seemed to be consistently followed by outright failure or at best equivocal results. Probably in the history of the disease the frustration in dealing with MS reached its pinnacle at this point. It was on this background that the use of interferon-β 1b emerged.

6.4 History of the Use of Interferon in MS

In the 1970s and early 1980s hypotheses on the cause of MS included a prominent role for viruses. Diseases causes by "slow" viruses had been described and the thought that a virus infection occurring early in life could result in a disease of young adulthood was attractive. In addition several pieces of evidence suggested a relationship between MS and common viruses of man such as measles virus. Thus, antiviral strategies including the use of interferon emerged as potential therapeutic strategies in MS. The first important study of interferon therapy was done by Jacobs et al. [13]. In this study, natural interferon-β was injected into the subarachnoid space of MS patients. The study, a two-arm study comparing the interferon treated patients to a group of untreated patients, reported a beneficial effect of interferon-β on relapse rate and progression. Following this report, a number of relatively small studies examining the effect of both β and alpha interferon were preformed with variable results. In 1990, a double blind, placebo controlled study comparing 2 doses of interferon-β 1b on exacerbation rate in patients with relapsing remitting MS was begun in North America [14]. This study, completed in 1993 reported a significant reduction in exacerbation rate in the patients treated with the higher dose of interferon. In addition to a

34% reduction in exacerbations, the patients treated with the high dose had a significant delay in the time to the first exacerbation while on treatment and had a reduction in the courses of corticosteroids and hospitalizations. The effect on progression was examined as a secondary outcome and while a reduction was seen in the high dose group, it did not reach statistical significance.

The North American trial of interferon-β 1b was distinguished not only for demonstrating a clinical effect but also for being the first important clinical study using MRI as an outcome measure [15]. The extend of disease as measured by the amount of brain with increased signal on T2-weighted images (lesion load) was studied yearly and a significant reduction in lesion load was seen in the treated patients with the greatest reduction in the patients receiving the high dose. The confirmatory MRI results played an important role in the approval of this treatment for MS by the Federal Drug Administration.

Although the effect on exacerbation rate was modest, the study and subsequent approval of interferon-β 1b resulted in the first treatment for RRMS proven to be at least partially effective and provided physicians and patients their first realistic treatment option for patients with RRMS. In distinction to the face value of these results other, subtler, results emerged. The often common practice on both the part of the physician and patient to deal with early MS through denial was challenged by the availability of a medical therapy. Thus, the attitude of the physician caring for patients with MS has had to be adjusted to the new problems brought by this treatment. When should a patient be started? Should treatment be discussed with patients with very early MS or should this discussion be held until some evidence of accumulation of disability was seen? How long should a patient be treated and how could the physician decide if the treatment was not effective in an individual patient. These questions probably have no absolute right or wrong answer and they are still in the process of being worked through by many physicians. Some saw the results of the clinical study encouraging and have urged patients to start treatment early. Other physicians have taken the position that the effect of treatment was so modest that it was barely worth the cost. In fact, this later position has been taken by health agencies in some countries leading to regional differences in the use of the treatment. Overall the approval of interferon-β 1b has changed forever the mind set of physicians caring for MS patients bringing both a new optimism as well as new frustrations.

Another important effect of the study and approval of interferon-β 1b was the demonstration that a treatment could be tested in MS and shown to have a significant treatment effect that would be sufficient to lead to approval by the FDA. Prior to this event some investigators including some in the pharmaceutical industry wondered if any treatment would ever show an effect in MS sufficient for approval of the therapy. There was concern that the disease was so variable and heterogeneous that proving a treatment effect might be impossible. Clearly, this was wrong and the approval of interferon-β 1b has lead to a new enthusiasm for testing new therapies in MS. Over the past few years numerous clinical trials testing new therapies have been initiated. One of these, a study of glatiramer acetate (copolymer-1), was published in 1995 and resulted in the approval of Copaxone for the treatment of RRMS [16]. At the time of approval of interferon-β 1b, a trial of another β interferon, interferon-β 1a was already underway. The study of interferon-β 1a differed from the study of interferon-β 1b in that, in distinction to examining exacerbation rate, the primary outcome measure was progression of disability. The study, published in 1995, demonstrated a significant reduction in progression as well as a reduction in exacerbation rate that was similar to that shown with interferon-β 1b [17]. Interferon-β 1a was approved for use in MS in 1996. Now for the first time physicians and patients had a choice of therapies for RRMS; each drug has its own advantages and disadvantages and the use of these treatments have been the subject of several recent reviews [18,19]. Most important, what do these treatments really do to the evolution of the MS lesion?

6.5 Inteferon-β and the MS Lesion

The results of the pivotal trial of interferon-β 1b indicated that treatment reduced exacerbations and slowed the increase in lesion load on MRI. The study did not provide much insight into the mechanism of the treatment effect. One important question was if the treatment affected the early stage of lesion development which can be measured using contrast enhanced MRI. A suggestion existed that this could be the case since a subset of patients in the pivotal study followed every 6 weeks

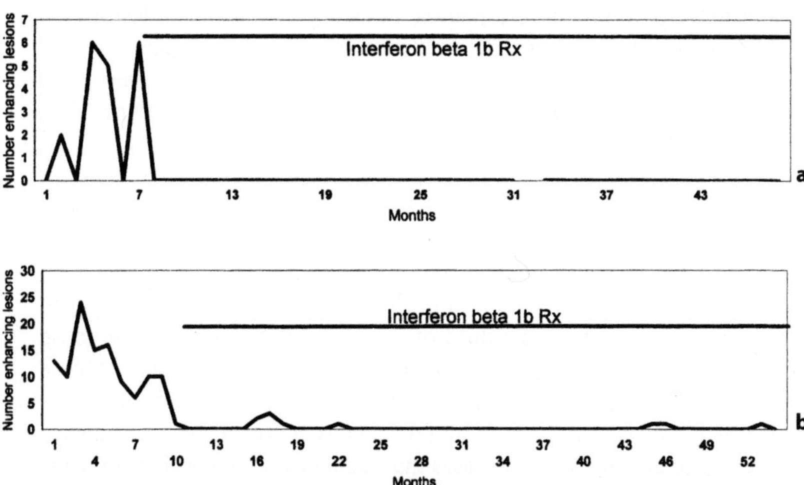

Fig. 2a,b. Effect of interferon-β 1b on the frequency of contrast-enhancing lesions. *Rx*, treatment

with T2-weighted MRI seemed to show a reduction in the number of new lesions. To answer this question, a trial was designed in 1993 to examine the effect of interferon-β 1b on contrast-enhancing lesions [20]. This study used a baseline versus treatment or cross-over design in distinction to the more common parallel groups design in which a treatment is compared to a placebo arm. The cross-over design had been reported to be an attractive approach for examining the effect of a treatment on contrast-enhancing lesions and had the advantage of requiring relatively small numbers of subjects [7,21]. If patients are studied monthly for several months, the average level of enhancing activity can be ascertained with reasonable accuracy. Since the cross-over design tends to eliminate the large interpatient variability in lesion frequency and since the serial baseline tends to reduce the affect of the large interpatient variability from month to month, the sample size needed to show a treatment effect is markedly reduced from that needed in a parallel groups design. Thus, a cohort of 14 patients was initially studied. Six- monthly baseline MRIs were done prior to beginning treatment followed by 6-monthly MRIs on treatment. The results indicated that interferon-β 1b produced a marked reduction in enhancing lesions; an 84% reduction was seen. The reduction in enhancing lesions occurred rapidly with a reduction occurring by the first month after beginning treatment. An example of the response in two patients is shown in Fig. 2.

Because of the extensive baseline analysis done on each patient prior to starting treatment, the effect of treatment could be examined in individual patients. The study cohort was expanded to a total of 33 patients and the findings indicated that when patients are followed for a year or more on treatment, different patterns of response begin to emerge [22]. In about one-third of the patients the treatment response with respect to reduction of enhancing lesions persists; in many of these patients new activity is completely eliminated. However, there is another group of patients that have an initial response to treatment but then have a return of activity. In some cases the frequency of new lesions remains below what was seen during the baseline period and in other cases activity returns to baseline levels. While some of the patients having a return of activity also have neutralizing antibody, the relationship seems complex and the explanation for loss of response is unclear in some patients. These results do provide additional insight into the effect of interferon treatment of MS. They indicate that the modest reduction in exacerbation frequency seen in the pivotal trial may reflect a heterogeneous response and not a partial or incomplete response in all patients. If correct this interpretation would add yet another dilemma for the physician and patient; how does one know in the care of an individual patient if the treatment is effective? Without the advantage of serial imaging available in a research setting, the question of efficacy in individual patients cannot be answered with certainty using MRI. Thus, the physician must base decisions concerning the effect of the therapy on the clinical course in the patient. Patients with continued clinical activity on treatment will easily be identified as treatment failures. What will not be identified are those patients who are clinically stable but with a frequent occurrence of contrast-enhancing lesions.

Over the past couple of years the value of interferon treatment in MS has been confirmed by additional studies. As mentioned above, the positive results obtained with interferon-β 1b were confirmed and extended by the study of interferon-β 1a given IM [17]. More recently the results of a study of interferon-β 1a given SQ have shown a reduction in exacerbations and a suggestion that the treatment effect may be dose-related [23]. Also just reported has been the study of interferon-β 1b in secondary progressive MS []. In this study treatment produced a significant delay in progression of disability. Thus, there is no longer doubt that treatment with interferon-β reduces the frequency of attacks in MS and has a modest affect on progression but important questions remain. Included are questions regarding patient selection for treatment, identification in practice of patients not responding and the relative merits of different doses and routes of administration. Despite the persistence of questions the identification of a treatment for MS, even if it produces only a partial reduction in exacerbation frequency and modest slowing of progression, has provided new hope for patients and an important tool for the physician. Overall the historical significance of the identification of a treatment effect with interferon-β and the approval of this drug for use in MS has brought both the care of MS patients and research into the disease into a new era. Much of the previous pessimism and discouragement felt in dealing with the illness has disappeared. Physicians are now able to take a proactive approach in dealing with the illness and it turn can provide new hope for patients.

References

1. Weinshenker BG (1994) Natural history of multiple sclerosis. Ann Neurol 36 [Suppl]: S6–11
2. McFarland HF (1998) The lesion in multiple sclerosis: clinical, pathological, and magnetic resonance imaging considerations. J Neurol Neurosurg Psychiatry 64 [Suppl 1]: S26–30
3. Frank JA et al (1994) Serial contrast-enhanced magnetic resonance imaging in patients with early relapsing-remitting multiple sclerosis: implications for treatment trials. Ann Neurol 36 [Suppl]: S86–90
4. Molyneux PD et al (1998) Correlations between monthly enhanced MRI lesion rate and changes in T2 lesion volume in multiple sclerosis. Ann Neurol 43:332–339
5. Stone LA et al (1995) Blood-brain barrier disruption on contrast-enhanced MRI in patients with mild relapsing-remitting multiple sclerosis: relationship to course, gender, and age. Neurology 45:1122–1126
6. Thorpe JW et al (1996) Serial gadolinium-enhanced MRI of the brain and spinal cord in early relapsing-remitting multiple sclerosis. Neurology 46:373–378

7. McFarland HF et al (1992) Using gadolinium-enhanced magnetic resonance imaging lesions to monitor disease activity in multiple sclerosis. Ann Neurol 32:758–766

8. Miller DH et al (1998) The role of magnetic resonance techniques in understanding and managing multiple sclerosis. Brain 121:3–24

9. Filippi M et al (1994) Quantitative brain MRI lesion load predicts the course of clinically isolated syndromes suggestive of multiple sclerosis. Neurology 44:635–641

10. O'Riordan J et al (1998) The prognostic value of brain MRI in clinically isolated syndromes of the CNS. A 10-year follow-up. Brain 121:495–503

11. Optic Neuritis Study Group (1997) The 5-year risk of MS after optic neuritis. Experience of the optic neuritis treatment trial. Optic Neuritis Study Group [see comments]. Neurology 49:1404–1413

12. Smith ME et al (1993) Clinical worsening in multiple sclerosis is associated with increased frequency and area of gadopentetate dimeglumine-enhancing magnetic resonance imaging lesions. Ann Neurol 33:480–489

13. Jacobs L, O'Malley J, Freedman A, Ekes R (1981) Intratrhecal interferon reduces exacerbations in multiple sclerosis. Science 214:1026–1028

14. The IFNB Multiple Sclerosis Study Group (1993) Interferon beta-1b is effective in relapsing-remitting multiple sclerosis. I. Clinical results of a multicenter, randomized, double-blind, placebo-controlled trial. Neurology 43:655–661

15. Paty DW, Li DK (1993) Interferon beta-1b is effective in relapsing-remitting multiple sclerosis. II. MRI analysis results of a multicenter, randomized, double-blind, placebo-controlled trial. UBC MS/MRI Study Group and the IFNB Multiple Sclerosis Study Group [see comments]. Neurology 43:662–667

16. Johnson K et al (1995) Glatiramer acetate reduces relapse rate and improves disability in relapsing-remitting multiple sclerosis: results of a phase III multi-center, double blind, placebo controlled trial. Neurology 45:1266–1276

17. Jacobs LD et al (1996) Intramuscular interferon beta-1a for disease progression in relapsing multiple sclerosis. The Multiple Sclerosis Collaborative Research Group (MSCRG) [see comments; published erratum appears in Ann Neurol 1996; 40(3):480]. Ann Neurol 39:285–294

18. Martin R, McFarland H (1996) Experimental immunotherapies for multiple sclerosis. Springer Semin Immunopathol 9:1–24

19. Thompson A, Noseworthy J (1996) New treatments for multiple sclerosis: a clinical perspective. Curr Opin Neurol 9:187–198

20. Stone LA et al (1995) The effect of interferon-beta on blood-brain barrier disruptions demonstrated by contrast-enhanced magnetic resonance imaging in relapsing-remitting multiple sclerosis. Ann Neurol 37:611–619

21. Miller DH (1996) Guidelines for MRI monitoring of the treatment of multiple sclerosis: recommendations of the US Multiple Sclerosis Society's task force. Mult Scler 1:335–338
22. Stone LA et al (1997) Characterization of MRI response to treatment with interferon beta-1b: contrast-enhancing MRI lesion frequency as a primary outcome measure. Neurology 49:862–869
23. Jacobs LD et al (1995) A phase III trial of intramuscular recombinant interferon beta as treatment for exacerbating-remitting multiple sclerosis: design and conduct of study and baseline characteristics of patients. Multiple Sclerosis Collaborative Research Group (MSCRG). Mult Scler 1:118–135
24. PRISMS (1998) Randomised double-blind placebo-controlled study of interferon beta-1a in relapsing/remitting multiple sclerosis. PRISMS (Prevention of Relapses and Disability by Interferon beta-1a Subcutaneously in Multiple Sclerosis) Study Group. Lancet 352:1498–1504
25. European Study Group on Interferon beta-1b in Secondary Progressive MS (1998) Placebo-controlled multicenter randomized trial of interferon beta-1b in treatment of secondary progressive multiple sclerosis. Lancet 352:1491–1497

7 Interferon:
The Dawn of Recombinant Protein Drugs
Closing Discussion

Dr. Jean Lindenmann: I have recently read a book by Jan Golinski who is a historian of science trying to explain to his graduate students what modern history of science should be. In his introduction he writes: "The history of science has had a long struggle to free itself from science's own view of its past." This may explain to you why there can be a discrepancy between what eye-witnesses or participants in certain developments believe to have lived through and what an outsider who looks at things from a much more general perspective may perceive.

So these two views of the history of science are not entirely congruent, but I feel that it is important that practitioners of science and real historians of science should at least communicate their ideas.

One of the most profound challenges in our traditional view of how science operates has come from Professor Kawade. I want to start this discussion by asking him one question, although I know that he does not like to be asked questions. Obviously, we will have to read his text very carefully before we can talk about it. But this is the naïve question which I would like to ask: You have criticised or discussed the reductionist view of science. You have opposed to it the biosemiotic view. The reductionist at least knows what he has to do. He will study molecules and then interactions between molecules and then networks and things become always more and more complicated. Maybe he will never reach a complete understanding, but at least he knows how to occupy himself. What does the biosemiotician do except waiting for the reductionist to give him the data?

Prof. Yoshimi Kawade: I am not against reductionism. That is a necessary method in science. But I am against the view that reduction is the only reliable, the only objective way of doing science. There are other ways I think. Reduction is certainly the most powerful method of approaching the world. That is necessary, but I am against the view that it is the only reliable way.

Dr. Derek Burke: Just to respond to the last two points: Of course, reductionism has been a very powerful method in our lifetime. But like Dr. Kawade, I also have concerns about its continued extension. We have a very active debate running in Great Britain at the moment where a Professor named Richard Dawkins in Oxford is basically saying that everything is explained by DNA, by genes, by the way in which genes interact with each other, and there is nothing else.

My objection to reductionism is when it shows itself as offering a complete explanation, e.g., when it says that all that we experience, know and study is nothing but the action of genes. My analogy is that this is nothing but a black box on white paper, but it carries meaning at a different level. Where I think the danger comes up in the applications of medicine is that we cannot treat the patient as merely a machine. This has come up several times this afternoon. Over and over again, in modern medicine we are discussing whether it is ethical to withhold treatment, what ethics govern what we should tell the patient, what ethics govern the use of new technologies. Those ethical values are not found inside science. They come in from elsewhere in our society.

So my concern is perhaps a slightly different one from Dr. Kawade's, that – impressed as we are by the effectiveness of reductionism – we think there is no other way to guide our society. I think in that way lies danger.

Prof. Otto Haller: I would like to comment on the first studies of interferon therapy. This morning, studies in which interferon was used have been listed. I missed something on that list. If I remember correctly, one of the first studies with interferon that proved to be of some promise was actually against herpes simplex virus in infections of the eye. These studies were performed by Sundmacher and Neumann-Haefelin in Freiburg, in a collaboration with Kari Cantell. They applied ointments or drops and showed that interferon had some effect. I think that was quite important at the time. This treatment became less important when acyclovir became available, and interferon is not in use anymore.

Dr. Ion Gresser: You usually combine it with TFT.

Dr. Kari Cantell: I have very pleasant memories of these studies. They began first with the British colleagues Perry Jones and Norman Finter, but most of the studies were really carried out in this country, in Freiburg, by Dr. Rainer Sundmacher and Dieter Neumann-Haefelin with interferon supplied from Finland for their studies.

The studies received very little publicity, unlike the cancer studies at the same time, but they made progress in small steps. There were many disappointments and for example, it turned out that if you use interferon drops, even very potent interferon, on its own it is completely ineffective. But when you combine this in itself ineffective interferon with a suitable therapeutic partner like acyclovir, then this combination is highly effective. I think this is a remarkable thing, that interferon, ineffective on its own, can be effective in a combination therapy. This was really the first hint how interferon should be used.

Now, as we have heard today, combination therapy is probably going to be the principle how interferons will be used in the future for the treatment of tumors, infections and probably also MS, I bet.

Dr. Ion Gresser: How do you know in those studies that it was the combination that made the interferon more effective rather than the interferon making the TFT more effective?

It reminds me of a study that I think Susan Krown reported, in which they used radiation to treat Kaposi's sarcoma of the mouth. The patient had received interferon. They found that there was a very marked enhancing effect of interferon on the radiation-induced tissue necrosis. They found that even the tissues which normally would not have been affected by that dose of radiation were severely damaged in this patient who had been receiving interferon.

Dr. Kari Cantell: You are quite right. My wording was wrong. I don't know whether interferon made TFT more active or vice versa. I am trying to say that the combination was more effective than either agent on its own.

Dr. Ion Gresser: Would you comment on the use of interferon rendering X-rays more effective? You know probably about that.

Dr. Kari Cantell: I don't know enough to comment on that.

Dr. Toine Pieters: I was curious what came of the studies done by Falkoff in South America dealing with visceral leishmaniosis. The photos he showed me looked liked a wondrous affair because things disappeared, the most obvious clinical effects I have ever seen dealing with interferon. Can somebody comment on that?

Dr. Ion Gresser: I think it was a combination of interferon-gamma and antimonials, wasn't it? I think the results really were very impressive.

Dr. Norman Finter: There was a study in Kuala Lumpur 10 - 15 years ago in which lymphoblastoid interferon was given to patients awaiting radiotherapy for inoperable naso-pharyngeal carcinoma. There was clear evidence that the interferon treatment potentiated the effect of the radiation. I simply don't recall the details or if they were published, but this is another example of where interferon acted as a radiation sensitizer.

Prof. Otto Haller: I would like to come back to multiple sclerosis for a moment. Everyone, of course, asks: How does interferon-beta work? The answer is: We don't know. As far as I understand, that's probably the correct answer at the moment. For the time being, the idea is to show that interferon has some clinical effect. I think there is good evidence at the moment, but it needs to be explored further.

But if interferon works, one really ought to be able to tell the patients how it works. Given that interferon-beta is such a pleiotropic molecule and the disease is not well understood, I wonder, whether there are good proposals to find out how interferon-beta works. There is no animal model. How would one go ahead?

Dr. Henry F. McFarland: I absolutely agree that this is a critical question. I think the problem has been the one that you ended up with, that there is no good animal model and in those that are used there are certainly issues of specificity with interferon. So, even examining the immunological mechanisms in animal models becomes, I think, problematic.

The studies in patients, in humans receiving interferon haven't been as extensively studied as they should have been. But it has been a very difficult thing to study. Generally, what has been studied are changes in peripheral blood, immunological parameters that tend to correlate with interferon treatment. There are changes, but they don't really correlate very well with the therapeutic effect.

In addition, there are a lot of in-vitro studies, but extrapolating those in-vitro findings into the in-vivo situation or into the organism has been difficult and so it has remained sort of a black box, I am afraid.

I think that there is a fundamental lack of understanding of the disease process. Clearly interferon seems to affect a very early stage in the lesion development. It seems to affect that stage that is characterised by opening the blood-brain barrier. We are not absolutely sure what causes that. We think, based on the animal models, that it's migration of an activated T-cell into the brain.

But if you step backwards for a second and look at the MRIs I showed, with gadolinium enhancing lesions, and you look at the number of lesions – some of the patients can have 10 or 20 enhancing lesions a month – and you try to put that together with an immunological explanation, it becomes a little bit difficult. You have to ask why the frequency of these focal lesions changes so much from month to month and why they seem to occur with a random distribution in the brain. One explanation immunologically is that you are doing what one might call an in-vivo limiting dilution experiment. So the more activated T-cells there are, the more likely you are to have a lesion and it's sort of a random hit event. However, immunological studies have not shown differences in the frequency of T-cells reacting with myelin that are consistent with this explanation.

But I think there is probably a more fundamental question. That is: does the process really start on the peripheral blood side or does it start on the brain side? Certainly, from a biological standpoint, one could probably make a very nice hypothesis that if you had a persistent virus that could become periodically activated and could produce a focal change in the blood-brain barrier, that could be the first event. Is it possible that interferon could target an event like that? Obviously, the answer would be yes. Although I think the mechanism is immunologically mediated, I am not sure we know it with 100% certainty.

Prof. Otto Haller: Previous reports have shown that interferon is present in these lesions. These finding supported the hypothesis that there may be a virus involved. There is already interferon present in the patient. Yet, with giving more interferon you have these beneficial effects. How does all this come together?

Dr. Henry F. McFarland: I think it's even more complex. If you look at these lesions from an immuno-histochemical standpoint, one finds many different cytokines both Th1- and Th2-like. From an immunological standpoint, we like to think of this Th1/Th2 paradigm, that Th1 cells are bad and Th2 cells are good and they produce regulatory cytokines such as IL10 and IL4 or TGF-beta. But if you look at one of these lesions by immuno-histochemistry, you find all the cytokines. Not only do you find all the proinflammatory cytokines, even in the very early stages, but you see all the regulatory cytokines. So it doesn't really fit very well, this yin-yang hypothesis. I agree with you.

Dr. Ion Gresser: One way might be to find out what interferon does in other demyelinating diseases. Maybe this would give you a clue. Sort of extending from the work with multiple sclerosis, what has been done with the use of interferon in other demyelinating diseases?

Dr. Henry F. McFarland: Other demyelinating diseases such as?

Dr. Ion Gresser: Lateral sclerosis.

Dr. Henry F. McFarland: Are you talking about ALS? That's a neuronal disease. There are others. Schilder's is probably a variant of MS. PML may have some demyelination associated with it, certainly subacute sclerosing panencephalitis has some demyelination associated with it. It's interesting, SSPE, as you know, has been treated with alpha-interferon, with, at best, mixed results, as far as I can see.

Dr. Kari Cantell: That's right. But these were just single cases and it was difficult to draw definite conclusions from those studies.

Dr. Henry F. McFarland: This is actually an interesting point because there are still people that treat their SSPE patients by infusing alpha-interferon. One wonders whether beta-interferon might be a better choice, if there is a component of an immunological mechanism in SSPE. So if one wants both an anti-viral effect and an anti-immunological, an immunosuppressive effect, beta-interferon might be a much better choice than alpha-interferon. I don't know of any SSPE patient that has ever been treated with beta-interferon. It might be an interesting study to consider at some point.

(Dr. Kapp: I just have a brief question going back to what you asked, Dr. Haller, because it seems to me from your question that you assume that beta-interferon needs to access the brain lesion? – Okay, if it doesn't, I don't need to ask my question.)

Dr. Ilana Löwy-Zelmanowicz: I just wanted to continue on the question Professor Burke posed: What does it mean to treat the patient? Dr. McFarland, you are telling us that MRIs are pretty secure in prediction in an early stage of the disease. What type of person who arrives with early signs of multiple sclerosis – one of the functions of the doctor is also to give good news and not bad news -, what is the percentage of patients who can be told with certainty: You will be fine, you will have no big problem, probably in 20 years, you will be in the good group?

The second question is: for those who come with mild signs of the disease, a first attack without any following events, and who are told that their brain activity, plaques etc. is bad and they are in a bad prognosis group and you recommend strongly I guess – doctors are quite persuasive to tell their patients, you should get treated because it's pretty sure that without treatment your situation may be much worse -, what are the consequences for the patient on the level of secondary effects, consequences for lifestyle? Also financial consequences may be a problem, will insurance take it on etc.?

Dr. Henry F. McFarland: I guess the first part of the question is: what is the data that MRI has any predictive validity? Probably the best data comes from a study that was started about 10 years ago at Queens Square. In this study David Miller and Allan Thomson and their colleagues took patients that were presenting with what is called monosymptomatic or clinically isolated syndromes. These were patients that were presenting with what would represent a first initial attack of MS. They imaged those patients and they brought the patients first back in 5 years. Then they asked the question: how many of those patients had gone on to develop clinically definite MS? For clinically definite MS you need to have two attacks, separated in time, and two areas in the nervous system involved. What was the relationship between the number of patients who developed clinically definite MS and what their MRI showed on presentation?

This paper – the first author was Massimo Filippi – reported that the patients that on presentation had a lesion load on T2 of 1.2 cc – that would represent three or four of these areas of increased signal – had about a 90% chance of developing clinically definite MS over 5 years. They had about a 50% chance of having an EDSS score of 3 or greater. The patients that had normal MRIs at presentation had a 6% chance of developing clinically definite MS.

That study has now been carried out for 10 years and was just published in *Brain*. The numbers are almost exactly the same. I think it was 85% for the patients with abnormal MRIs and about 10% clinically definite MS for patients that had normal MRIs. So if one looks at that very early stage, clearly what you see on an MRI does have a fair degree of predictive value. Once you get beyond that, once you get patients that are further on in the disease, to take any MRI parameter and try to predict forward becomes more difficult, partly because we don't have a lot of good data.

If you try to do cross-sectional studies, the correlations are really poor, but that's not really what you want to do. You want to be able to predict. One of the efforts right now in the MS community is to try to gather together all of the existing data that has come out of the large number of clinical trials that have been done over the last several years and to put those in one simple data base so that we can try to look at these questions. Does that answer your question?

Dr. Ilana Löwy-Zelmanowicz: Being completely out of this area, I just wanted to know, to put it very simply: How many patients among those who are going to see a doctor with early MS are going to be told: You will be all right?

Dr. Derek Burke: She wants the ethical issues that lie behind your decision.

Dr. Henry F. McFarland: I will give you more data because you asked for a percentage. We have a group of 150 patients that we have done 3-monthly MRIs. Out of that 150 patients, 70% will have at least one enhancing MRI over that 3-month period. So if I take 100 patients with early relapsing-remitting MS, about 70% of them over 3 months will have some activity. On a single MRI it numbers about 50%. So about half. So the answer is that if you see patients early, about half will have an MRI that if I looked at it, I would say: I think you should consider treatment.

Now we get into the ethical issues. I think we talked about that a bit before. I think they probably vary from one physician to the other. My personal feeling is that patients should be given as much power and control over their own lives as we possibly can. I think one shouldn't dictate. The way I handle the situation is to ask the patient if they want to know what their MRI shows. Most will say yes. But there are a few that say: I don't want to know. I will say: do you want to discuss the implications for treatment? Some will say no. That's a very small number, but if they say no, that will be the end of it.

If they said, I want to discuss it, I will try to give them the same information that I am giving you, probably in a way that they can put it into perspective. The next part of your question is: what is the effect of treatment on their lifestyles? I think it is relatively small with these treatments. There is some impact, there is no question about that.

I think a whole other issue one has to address is: What are the implications of diagnosing the disease in a young individual? I think one has to be very careful of that because there are implications for insurance, there are implications for employment. This will vary geographically. There are social implications. For example, a lot of the patients that I see are young people who may be engaged for example. So you have a young woman walk in who is engaged to be married in 6 months. And you tell her that she has MS. People will deal with that differently. Be it a male or a female, that will have implications on whether they want to get married. So you bring up a lot of issues. I think the clinician has to be able to deal with that.

I can't speak for Europe, but in the US there probably are some problems with managed care, that physicians have less and less time to spend with patients. So on the one hand you have very powerful diagnostic tools, you diagnose the disease very early, but the ability to support the patient after you give the diagnosis often times is not there. That's unfortunate.

Dr. Ilana Löwy-Zelmanowicz: Usually, the interferon therapy is covered by usual insurance or not?

Dr. Henry F. McFarland: It is almost always covered. Depending on the insurance, patients sometimes will have to pay small amounts. All of the pharmaceutical companies will under certain circumstances provide their medication essentially free to people that have no coverage. So people that have true need will get their medication from the pharmaceutical companies. Pharmaceutical companies have been very good about that actually.

Dr. Wolf-Dieter Schleuning: I would like to touch an issue which hasn't been considered during the whole meeting. That is the issue of the genetics of multiple sclerosis. We all experience a rapid increase in knowledge of the human genome. We have now achieved very high marker density. It is now in the realm of reality that we will very soon be able to make whole genome scans at reasonable cost. There are so far about three quite well-defined predisposition loci for multiple sclerosis.

It is quite clear from epidemiology that there is a genetic contribution to the disease phenotype. I would just like to know from you: How will this in the future impact the way patients will be treated? Do you think that it will be relatively soon a reality that there will be genetic stratification of patient groups? Are there already any data that there may be a genetic disposition to a response or non-response, e.g., to interferon? So, how do you see the future of that field?

Dr. Henry F. McFarland: I guess I would probably first look at the data on genetics a little more pessimistic than you presented it. There is absolutely no question that there is a very strong genetic influence on susceptibility. Look at the two case histories that I had on slides today. Both of those patients had family histories of MS. The concordance rate is high on monozygotic twins, probably approaches 50%. So there is no question that there is a genetic influence.

On the other hand, if one really looks at the three major genome scan studies – the study from the UK, the study from Canada, the study from the US – the results to my mind are a bit disappointing. Two of them identified MHC, one partially identified MHC. If you get beyond that, the data get a bit soft.

Obviously, we are dealing with a complex disease. There are probably multiple genes that influence susceptibility. I think it is probably equivalent to what we see in the animal models, the diabetic, NOD mouse or the lupus glomerulonephritis model. I think in both of those models one sees multiple genes that can influence susceptibility. Depending on how many of those genes you have, influences your risk of getting the disease. But you can come up with similar risks whether you have genes A, B, and C, or A, C, and D. So, if you extrapolate that into doing studies in heterogeneous populations, such as exist in the United States and generally exist in Europe, I think, identifying susceptibility genes probably is going to be very difficult.

Look at the Finnish data. It gives a clearer picture. For example, there is a gene that is an upstream myelin basic protein, a gene that looks like it affects susceptibility: that has not been identified in any of the other genome screen studies. But what that would suggest is: in that population, there may be a different gene coding for susceptibility, which may in part account for some of the differences in the frequency or incidents of disease in different populations. There is probably another gene in the Finnish study, that also has an influence. I believe those are the genes you were thinking about.

So, I think it is going to be very complicated. And I think the likelihood of very soon getting the type of information that would ever be used for genetic counselling will be a long way down the road.

On the other hand, the other issue you bring up is: to what extent can we generate genetic data that will help us in looking at subpopulations of patients? That to me is a very exciting area. Clearly, if you just look at the interferon data. There are differences. Patients behave differently with respect to their clinical courses, their MRIs behave differently.

I think using biological markers – whether they be genetic or immunological markers or whatever – to try to stratify these groups of patients is clearly an area that really needs to be focused on. Then I think we will move much more quickly.

Prof. Patricia Fitzgerald-Bocarsly: I just wanted to throw out the question of whether the audience thinks that there is going to be a desire on the part of either federal funding agencies or pharmaceutical companies to really understand mechanisms and take it apart better how different cytokines are working or whether or not they can be used in combination or is it just too expensive and too complicated?

Dr. Henry F. McFarland: I don't know the answer to it, but speaking unofficially from NIH: the National Institute of Health now has created a new grant mechanism that is designed to fund proof of principle studies which I think is an exciting concept. So if one plans a clinical trial, whether it is supported by an NIH grant or a pharmaceutical company, often times within that funding for the clinical trial is not the money for proof of principle aspects or immunological studies or biological studies, whatever they might be. There is now a separate funding mechanism for those which is sort of a fast track. So I think that will help.

Dr. Norman Finter: The question of where we go in the future is very difficult to answer. There are probably at least 60 different cytokines, and possibly many more. If one thinks in terms of using combinations of cytokines, which ones should be chosen and on what basis?

I think there are two options. One is to go along with what happened in the early days of antibiotics. People screened material from hundreds of sources, hoping that something useful might result – indeed, at one stage almost every new soil sample yielded an antimicrobial agent, and some finally proved useful. At the same time, the large pharmaceutical companies were screening hundreds of thousands of novel organic compounds from their collections in the hope of finding an active antiviral substance. There were indeed some important successes from this approach, e.g. acyclovir and zidobudine, but the likelihood of success from this random approach is really very remote.

Perhaps in the future some new screening system will be devised – though I cannot at present see how this could be done – whereby combinations of cytokines can be tested in vitro or in vivo, and from the results, those more likely to prove clinically useful can be chosen.

An alternative approach is in my opinion much more likely to give results. This is to develop ways of measuring in real time and by as yet undevised methods the cytokines being formed in a particular part of the body in health and disease. Although this may seem very far-fetched, there are already some techniques which may make this possible, and one should never underestimate what will happen in the future. So I hope the results obtained in this way will provide a rational basis for developing cytokine mixtures which should correct the imbalance observed in a particular organ during the course of a disease.

As an example of something which could be tackled now, one could follow up Tom Merigan's studies in herpes zoster. These showed that the vesicle fluids contained relatively very high concentrations of interferons. By harvesting fluid from vesicles at different stages in their evolution, one could determine what are the predominant cytokines made at each stage as the cells accumulating at the base of the vesicle change from initially being predominantly polymorphonuclear neutrophils to different categories of B and T lymphocytes, and finally to monocytes. One would expect this to be mirrored by a gradually changing pattern of cytokines in the vesicle fluids.

I would like to take up a point already made by Yoshimi Kawade in his talk. The human body has acquired an almost incredible number of highly developed homeostatic mechanisms. When we try to use a cytokine or even a combination of cytokines to correct a temporary imbalance in disease, it is as if we were trying to regulate an expensive watch with only a single tool, capable of changing only a few functions. We can make an adjustment in good faith and hope that it will improve the time-keeping, but it is just as likely that most of our attempts will make matters much worse.

So I do hope that those working on the fundamental mechanisms of cellular behaviour and control will give us intelligent leads to follow, rather than that we shall merely blunder on, and devote huge resources to the empirical illogical and random approach.

We are in an area of "big" biological research. The pharmaceutical companies must aim for products which will have the huge sales and profits needed to sustain all the other work they carry out. The big independent or government research institutes usually focus their attention on a selected small number of targets, and are very slow and reluctant to change these. For example, most work of the U.S. National Cancer Institute in the 1970s was devoted to the search for anticancer chemical chemotherapeutic agents. It was only as the result of pressure from the powerful lobby organised by Mathilde Krim that the NCI finally started to devote staff and money to work on interferons and other biological topics.

I hope that vested interests, whether financial or other, don't take too much control, and bench scientists will be given the freedom and funds to carry out more speculative work. Some of their results are likely to point the way to more rational ways of treating patients with cytokines.

Dr. Ursula Lachnit-Fixson: A question to Dr. McFarland, touching a very practical problem, namely the prevention of pregnancy in young treated women. Do you allow the use of hormonal contraceptors or do you rather prefer non-hormonal methods for prevention of pregnancy?

Dr. Henry F. McFarland: Hormonal contraceptors I think are fine. There has even been some antidotal evidence that there may be some beneficial effects.

Dr. Ursula Lachnit-Fixson: In the early days of the oral contraceptors we included some remarks that in some cases we had even seen improvements.

Dr. Henry F. McFarland: Yes.

Dr. Jean Lindenmann: Thank you very much for your participation.

Subject Index

Ernst Schering Research Foundation Workshop

Editors: Günter Stock
Monika Lessl

Vol. 1 *(1991)*: Bioscience ⇋ Society – Workshop Report
Editors: D. J. Roy, B. E. Wynne, R. W. Old

Vol. 2 (1991): Round Table Discussion on Bioscience ⇋ Society
Editor: J. J. Cherfas

Vol. 3 (1991): Excitatory Amino Acids and Second Messenger Systems
Editors: V. I. Teichberg, L. Turski

Vol. 4 (1992): Spermatogenesis – Fertilization – Contraception
Editors: E. Nieschlag, U.-F. Habenicht

Vol. 5 (1992): Sex Steroids and the Cardiovascular System
Editors: P. Ramwell, G. Rubanyi, E. Schillinger

Vol. 6 (1993): Transgenic Animals as Model Systems for Human Diseases
Editors: E. F. Wagner, F. Theuring

Vol. 7 (1993): Basic Mechanisms Controlling Term and Preterm Birth
Editors: K. Chwalisz, R. E. Garfield

Vol. 8 (1994): Health Care 2010
Editors: C. Bezold, K. Knabner

Vol. 9 (1994): Sex Steroids and Bone
Editors: R. Ziegler, J. Pfeilschifter, M. Bräutigam

Vol. 10 (1994): Nongenotoxic Carcinogenesis
Editors: A. Cockburn, L. Smith

Vol. 11 (1994): Cell Culture in Pharmaceutical Research
Editors: N. E. Fusenig, H. Graf

Vol. 12 (1994): Interactions Between Adjuvants, Agrochemical
and Target Organisms
Editors: P. J. Holloway, R. T. Rees, D. Stock

Vol. 13 (1994): Assessment of the Use of Single Cytochrome
P450 Enzymes in Drug Research
Editors: M. R. Waterman, M. Hildebrand

Vol. 14 (1995): Apoptosis in Hormone-Dependent Cancers
Editors: M. Tenniswood, H. Michna

Vol. 15 (1995): Computer Aided Drug Design in Industrial Research
Editors: E. C. Herrmann, R. Franke

Vol. 16 (1995): Organ-Selective Actions of Steroid Hormones
Editors: D. T. Baird, G. Schütz, R. Krattenmacher